THE
Meteorite
& Tektite
COLLECTOR'S HANDBOOK

THE
Meteorite
& Tektite
COLLECTOR'S HANDBOOK

A Practical Guide to Their
Acquisition, Preservation and Display

Philip M. Bagnall

Published by:

Willmann–Bell, Inc.

P.O. Box 35025
Richmond, Virginia 23235 ☎ (804)
United States of America 320-7016

Publishers and Booksellers

Serving Astronomers Worldwide
Since 1973

Published by Willmann-Bell, Inc.
P.O. Box 35025, Richmond, Virginia 23235

Printed in the United States of America

Library of Congress Cataloging-in-Publication Data.

Bagnall, Philip M.
     The meteorite and tektite collector's handbook / Philip M.
    Bagnall
         p.   cm.
    Includes bibliographical references and index.
    ISBN 0-943396-31-X
     1. Meteorites– Catalogs and collections.   2. Tektites–Catalogs
and collections.   3. Astronomy–Amateurs' manuals.   4. Geology-
-Amateurs' manuals.   I. Title.
QB755.B24 1991   90-25854
523.5'1'075–dc20   CIP

91 92 93 94 95 96 97 98 9 8 7 6 5 4 3 2

Acknowledgements

In preparing this book I have sought the help of a number of individuals and institutions and I would like to record my deepest appreciation for their efforts. I would particularly like to thank Dr. Robert Hutchison of the British Museum (Natural History) who took the time and trouble to read and comment on the original draft manuscript. His criticisms and suggestions have greatly improved the text.

I am indebted to the following people for supplying photographs and advice and, in a number of cases, for their encouragement; Dr. Elizabeth Roemer, Dr. David W. Hughes, University of Sheffield; Harold B. Ridley, Dr. Brian Mason, Smithsonian Institution, Museum of Natural History; Dr. Arthur W. Struempler, Chadron State College; Dr. Vagn F. Buchwald, Technical University of Denmark; Dr. Colakoglu, Gazi University; Dr. John A. O'Keefe, NASA Goddard Space Flight Center; Dr. Peter M. Millman, National Research Council of Canada; Dr. P. R. Sen Gupta and Dr. Sibdas Ghosh, Director General, Geological Survey of India; Prof. G. Kurat, Naturhistorisches Museum Wien; Dr. Alex Bevan, Western Australian Museum; and last but, as they say, by no means least, my old friend James Shepherd.

The following institutions helped provide a number of photographs and permissions for which I am most grateful; Smithsonian Astrophysical Observatory, United States Geological Survey, NASA Jet Propulsion Laboratory, American Museum of Natural History and David & Charles (Publishers) Ltd.

Finally, this Handbook would not have been possible without the commitment of my wife, Pauline, and my son, Alexander, both of whom have shown considerable patience during the past few months.

To my son, Alexander

Table of Contents

Acknowledgements   iii

Introduction   ix

1 The Meteoritophiles   1

2 Sources of Meteorites   11
2.1 Introduction 11
2.2 Commercial Sources 11
2.3 Private Sources 12
2.4 Falls 14
2.5 The Recovery of Meteorites from Predicted Impact Sites . . 18
2.6 Finds 24
2.7 Craters and Crater Fields 26
2.8 Legal Ownership 32

3 Composition   33
3.1 Introduction 33
3.2 Isomorphism and Polymorphism 33
3.3 Minerals 38
3.4 Isotopes and Meteorite Ages 43
     3.4.1 Cosmic Rays 45
3.5 Do Meteorites Smell? — An Aside 47

4 Structure and Appearance   49
4.1 Introduction 49
4.2 External Features 49
     4.2.1 Fusion Crust 49
     4.2.2 Regmaglypts 54
     4.2.3 Oriented Meteorites 56
4.3 Internal Features 57

4.3.1	Breccias	58
4.3.2	Chondrules	58
4.3.3	Calcium- and Aluminum-Rich Inclusions (CAI's)	63
4.3.4	Neumann Lines	64
4.3.5	Widmanstätten Pattern	65
4.3.6	Reichenbach Lamellae	69
4.3.7	Stony-Iron Admixtures	69
4.3.8	Albedo	69

5 The Classification of Meteorites   **71**

5.1	Introduction	71
5.2	Modified Classification System	73
5.3	Stones	74
5.3.1	Chondrites—General	75
5.3.2	Enstatite Chondrites	80
5.3.3	Olivine-Bronzite Chondrites	81
5.3.4	Olivine-Hypersthene Chondrites	81
5.3.5	Amphoterites	81
5.3.6	Carbonaceous Chondrites	82
5.3.7	Achondrites—General	85
5.3.8	Angrite	85
5.3.9	Aubrites	86
5.3.10	Ureilites	86
5.3.11	Eucrites	86
5.3.12	Diogenites	87
5.3.13	Howardites	87
5.3.14	Shergottites	87
5.3.15	Nakhlites	88
5.3.16	Chassignite	89
5.4	Stony-Irons	89
5.4.1	Lodranites	89
5.4.2	Mesosiderites	89
5.4.3	Pallasites	91
5.4.4	Siderophyre	91
5.5	Irons	91
5.5.1	Hexahedrites	92
5.5.2	Octahedrites	93
5.5.3	Ataxites	93

6 The Finer Points of Meteorite Collecting   **95**
  6.1 Introduction 95
  6.2 The Identification of Meteorites 96
  6.3 Preservation 98
     6.3.1 Dust 98
     6.3.2 Clay 99
     6.3.3 Rust 99
     6.3.4 Deliquescence 99
     6.3.5 Light 100
     6.3.6 Cracking, Scratching and Fragmentation 100
     6.3.7 Lacquering 100
  6.4 Display Methods 101
     6.4.1 Cutting and Polishing 101
     6.4.2 Etching 102
     6.4.3 Heating 104
     6.4.4 Exhibiting Specimens 104
     6.4.5 Labelling 107
  6.5 Taking a Closer Look 108
     6.5.1 Meteorites Under a Hand Lens 108
     6.5.2 Microscopy 109
  6.6 Keeping Records 111
     6.6.1 Basic Details 111
     6.6.2 Photography 112

7 Tektites   **113**
  7.1 Introduction 113
  7.2 History 113
  7.3 Locations 114
  7.4 Appearance 116
  7.5 Structure and Composition 122
  7.6 Collecting Tektites 123
  7.7 The Origin of Tektites—An Aside 124

A Glossary   **127**

B Meteorite and Tektite Collections   **137**

C Useful Addresses   **141**
  C.1 Suppliers 141
  C.2 Organizations 142
  C.3 Nininger Award 143

D Bibliography   **145**

E Further Readings   149

General Index   151

Meteorite Index   157

Name Index   159

Introduction

Man has collected meteorites and tektites for thousands of years. They have been used as simple tools and weapons, revered and worshipped, discarded and displayed, and today they are the subject of intense investigation by some of the world's leading scientists. Once regarded as useless pieces of cosmic debris, meteorites, we now realize, hold clues vital to our understanding of the origin and evolution of the solar system. They are the cornerstone on which our current theories are based. But this book is not about what meteorites can tell us a subject that has been treated with admirable clarity by a number of authors recently—instead it is about the practicalities of collecting these fascinating objects. My objective in preparing this handbook is to advise the growing number of potential and experienced collectors on how to acquire, preserve, and exhibit specimens.

I have started this book with a brief review of the history of meteorite collecting. This is followed by details of how meteorites are priced and sold and three chapters dealing with composition, structure, and classification from a practical viewpoint. Chapter 6, "The Finer Points of Meteorite Collecting," deals with testing, preservation, and exhibition techniques and suggests ways in which the meteoritophile can derive the greatest rewards from collecting. The final chapter on tektites may seem slim, but many of the points concerning meteorites discussed in the preceding five chapters apply equally to tektites.

It has often been said that people who undertake systematic collections—be they of antiques, coins, stamps, or whatever—tend to undergo a transformation. What begins as a harmless past-time can rapidly evolve into an all-consuming passion in which the collector needs to know everything about his subject. Meteoritophiles are no exception. While some are quite happy to marvel at the beauty of meteoritic structure, others strive to learn about the composition and histories of the specimens they have in their collections. Furthermore, they are often willing to talk about meteorites and tektites to local astronomical and geological societies and to natural history groups. Although many professional meteoriticists

ix

welcome amateur interest in meteorites, not everyone believes in amateur involvement. They fear that a rare or unique specimen may find its way into the hands of a private collector, thus depriving science of what may be a vital piece of the cosmic jigsaw. Their views are easy to understand and not entirely without foundation. However, it should be remembered that some meteoritophiles have been very generous to the professional scientific community in the past, providing both specimens and funds for research. Of course, the way to allay the fears of the professional is to ensure that all new falls and finds are made available for analysis. The meteoritophile also has an important rôle to play in bringing meteorites to the attention of the general public in order to gain their support—especially in this era of meagre research grants!

My hope is that this handbook will assist those who already collect meteorites and tektites and will encourage those who have not already done so to embark on a fascinating and absorbing journey into the history of our planetary system.

P.M.B.

Chapter 1

The Meteoritophiles

Since the mid-1970's the number of amateur astronomers and geologists interested in collecting meteorites has grown steadily so that today there is a vibrant international trade in these fragments of cosmic débris. Meteorite collecting, however, is by no means a recent phenomenon, though the reasons for collecting meteorites have changed considerably over the centuries.

In Mexico and North America, Man had discovered meteoritic iron by the fourth millennium B.C. and had used it to make simple tools and weapons. Many of the early civilizations realized that meteorites fell from the sky—traditionally the home of the gods—and regarded them as heavenly gifts and signs of divine intervention in mortal matters. An inventory of treasure for one Hittite king includes among the jewels, gold, and silver, "iron from heaven," a clear indication of the importance which early civilizations attached to meteorites. The Greek philosophers Anaxagoras (500–428 B.C.) and Plutarch (A.D. 46–120) and the Romans Livy (59 B.C.–A.D. 17) and Pliny (A.D. 23–79) all mention meteorites in their writings, and in China, Mu Tuan Lin (1245–1325) recorded falls spanning more than two millennia. One European philosopher who openly disagreed with the view that meteorites were sent by the gods was Diogenes of Apollonia (412–322 B.C.), the Greek cynic who lived in a tub and who told Alexander the Great to get out of his sunlight. He regarded meteorites as "invisible stars that fell to Earth and died out, like the fiery stony star that fell to Earth near the Aegos-Potamos River." Although no one was willing to listen to him then, Diogenes is now remembered by having a class of meteorites—the diogenites—named after him.

The Egyptians were among several Middle Eastern civilizations who systematically collected meteorites. Hieroglyphics found on the interior walls of several pyramids tell of "heavenly iron" being interred within the tombs of the pharaohs. The Egyptians were, of course, firm believers in the after-life, and they would fill their burial chambers with food and wine so that they would neither thirst nor starve on their final inevitable journey. They also added treasures, pets, slaves, and even members of their own family in order to make their existence on the other side as pleasurable and as comfortable as possible. They obviously considered meteoritic iron to be of some considerable value.

Across the Mediterranean the Romans also regarded meteorites as sacred. A stone which fell in Phrygia was ceremoniously transferred to Rome in 204 B.C., and a meteorite that was discovered in the early years of the Roman Empire was presented to Numa Pompilius, the philosopher who reluctantly succeeded Romulus, the founder of Rome. He declared the shield-shaped iron to be vital to the security of the state and ordered eleven copies of it to be made in order to confound spies and saboteurs. Other meteorites which fell during the Roman era were placed in the Temple of Venus on the island of Cyprus. One such object, the Diopet or Zeus-Fallen Thing, eventually found its way into the headdress of the Statue of Diana at Ephesus. The Greeks were also great collectors of heavenly gifts and placed their finds in the Temple of Apollo.

Meteoritic iron has been used to make knives, nails, anvils, and (in Arabia) swords in the belief that their bearers would be invulnerable. In 1621 the Mogul Emperor Jehangir had a meteorite fashioned into a knife, a dagger, and a sword, and the Javanese Sultan of Solo had daggers made from the Prambanan iron (discovered in 1797), which he gave away as presents.

Built into the masonry at the Eastern corner of the Kaaba in Mecca is the Hadshar al Aswad or Black Stone, which some consider to be a meteorite. Mohammed (A.D. 570–632) was said to have wept when he touched the stone, and pilgrims to the shrine must touch and preferably kiss it. According to legend the stone was originally white but turned black because of the sins of Man. Although it has been known to Westerners since at least 1772, few non-Muslims have seen it. Those that have report that it is set in silver, is somewhat fragmented, and shows diffuse banding. This report has led critics to suggest that the rock may actually be agate.

Like many other early civilizations, the Japanese also collected meteorites as part of their religion, believing that they belonged to Shokujo, the Goddess of Household Skills, who used them to steady her loom. All Japanese meteorites were therefore placed in her temple.

The rise of Christianity virtually put an end to public and state interest in meteorites. For nearly one thousand years, from the time of the Venerable Bede in the eighth century, the Catholic Church denied the existence of rocks that had plunged from the firmament. Such objects, the Church maintained, were real only in the minds of the ignorant peasantry, and when events occurred that appeared to contradict the Church—as in the case of a well-observed fall—the Devil himself was blamed. A typical example was the 127 kg stony meteorite which fell at Ensisheim in Alsace on the morning of November 16, 1492 (Gregorian Calendar). The authorities declared it to be the work of the Devil and ordered it to be chained to the wall of the parish church so that it could not fly back to its satanic master. The locals were less than happy with this edict because they simply could not understand how the Devil in the bowels of the Earth could cast a stone from Heaven. Eventually the Church relented and decided that it must have been sent by God as a sign of His wrath. A peculiar twist to this tale is that only because of the actions of the oppressive Church regime was the Ensisheim meteorite preserved, and it is now the second oldest surviving specimen (the oldest fall is Nagata, Japan, 861).

The Inquisition actively discouraged the populace from showing an interest in meteorites. Anyone who disobeyed their wishes would be branded a heretic and could expect to be accused of dabbling in pagan cults and the black arts, the ultimate penalty for which was death. Under these circumstances, it is hardly surprising that interest in meteorites waned.

With the Enlightenment came confusion. The belief that those few meteorite falls acknowledged by the Church were "miracles" was rejected by scientists—but so too was the view that meteorites were from the depths of space. It has often been said that the scientists of the eighteenth century were still strongly influenced by religious dogma and superstition, but this is a misunderstanding by present-day writers of the situation that existed during the Enlightenment. The views of those early scientists were based on sound scientific reasoning. At that time it was generally believed that the planets and their satellites, together with the comets, were self-contained, discrete systems. In theory nothing existed between the planets other than an ill-defined aether, necessary to transmit light and heat from the Sun, and the force of gravity. Careful observation, moreover, failed to reveal any other objects—the first asteroid was not discovered until 1801. Since the time of Aristotle (383–322 B.C.), meteors and fireballs were considered to be meteorological phenomena and were regarded as "hot, dry exhalations which rose from the Earth and, carried by the atmosphere, became heated and burst into flames." When the nature of lightning was discovered a new method by which the "exhalations" could be ignited had been found, and because no one had actually seen a fireball drop a meteorite, the connection between these latter two phenomena could not made. The suggestion that meteorites themselves could form in the atmosphere was popular for a short time but was eventually dismissed when a greater understanding of mineralogy prohibited such processes. In addition, many of the "meteorites" that were presented to scientists as evidence turned out to be terrestrial rocks and Iron Age implements. Consequently, falls of rocks from the sky were placed in the same category as the fabulous rains of blood, milk, and frogs that were recorded by the early Greeks and Romans which scientists had rightly dismissed. Hence, meteorites were thought to be volcanic bombs and rocks that had been struck by lightning—a view that was so popular that meteorites became widely known as thunderstones. Abbé Stütz (1747–1806) was typical of eighteenth century researchers. Curator of the Austrian Imperial Cabinet, he described accounts of the Hraschina meteorite fall (May 26, 1751) as "fairy tales." Fortunately, he was sufficiently intrigued by thunderstones to preserve them—unlike some of his contemporaries—and his specimens later formed the basis of the now famous Vienna meteorite collection.

Several scientists in the mid-eighteenth century were convinced that there was more to thunderstones than most people believed. Among them was the Prague astronomer Josef Stepling (1716–1778) who, in 1753, witnessed the fall of several meteorites at Tábor in Bohemia. He took the view that the rocks had come from beyond the atmosphere, but one researcher went further. Father Domenico Troili (1722–1792) performed chemical analysis on the meteorite that fell at Albareto, Italy, in July 1766 and recorded an iron sulphide which is now called troilite in his honor. But the scientists who speculated on the cosmic nature of meteorites were lone voices and were largely ignored. It took an unusual series of events to eventually convince the world that meteorites were of extra-terrestrial origin.

In 1749 a 687 kg iron and stone meteorite was found in Siberia near Krasnojarsk. It was a magnificent rock with bright, silvery metal interspersed with rich yellow-green crystals of olivine and was soon recovered by a local blacksmith, Medvedyev, who thought "it was something better than iron." Some years later the Russian Academician Peter Simon Pallas (1741–1811) was charged by the Czar to explore the vast and largely unknown lands of Siberia. Quite by accident Pallas heard about the rock that Medvedyev had salvaged and went to investigate. He was just as impressed by the rock as the blacksmith had been, and had it moved five years later to the St. Petersburg Academy of Sciences. Once there sections were cut from the object and sent to museums and universities throughout the world. One piece was dispatched to Berlin where it caught the attention of the German physicist Ernst Florenz Friedrich Chladni (1756–1827). Chladni was fascinated by the specimen and spent hours examining it very closely. In 1794 he visited the main mass, which was still in St. Petersburg, and became convinced that it was cosmic in origin. He published his findings in German, at Riga, the capital of Latvia, under the title *On the Origin of the Iron Masses found by Pallas, and Others similar to it, and on some Natural*

Phenomena related to Them. In his publication Chladni not only proposed that meteorites came from space but that they were intimately associated with fireballs. Although one or two of his fellow scientists welcomed his report, most chose to ignore it—but not for long, for within a decade events were to prove Chladni right, and science underwent a paradigm shift.

On April 26, 1803, the small town of L'Aigle in Normandy, France, was pelted with between 2,000 and 3,000 meteorites, which left the townsfolk in no doubt as to the nature of the rocks. The French Academy of Sciences who were, almost to a man, naturally skeptical of "cosmic" objects sent their member Biot to investigate. His report of the fall, along with detailed chemical analyses, resulted in uproar at the Academy—but the truth could be dismissed no longer and the Academy was forced to admit that Chladni and his supporters were correct.

As news of the L'Aigle event spread throughout Europe and across the world, many museums and universities began to take a fresh look at their specimens and embarked upon a policy of acquiring new ones. Major collections were established in Vienna, most notably by Karl von Schreibers (1775–1852), Friedrich Mohs (1773–1839) and Paul Partsch (1791–1856), in London by Karl Konig (1774–1851) and Nevil Story-Maskelyne (1823–1911), and elsewhere. Specimens were catalogued and categorized, though early attempts at classification were not always successful. The Victorians were great collectors of curiosities—including minerals and fossils—and meteorites did not escape their attention. By the end of the nineteenth century, literally hundreds of meteorites had been placed in collections, both private and public.

The recognition of meteorites helped to clarify the nature of meteor and fireball phenomena which, in the 1800's, baffled many an eminent scientist. Researchers could not decide whether meteors were essentially meteorological processes, as Aristotle believed, or whether they resulted from an external influence. Meteorites settled this issue, although it was 1848 before James Joule suggested a mechanism by which meteorites could "glow" as they passed through the atmosphere.

Speculation soon began on the origin of meteorites. It was obvious that, in many ways, they were similar to terrestrial rocks, and so it seemed logical to assume that they were fragments of a disrupted planet. While updating the inaccurate star catalogue of Wollaston, Father Guiseppe Piazzi in Palermo, Italy, discovered a new object on January 1, 1801, that many believed was the "missing" planet which, according to the Titius-Bode Law, should have existed between Mars and Jupiter. However, by 1807 no fewer than four new "planets" had been found, all at approximately the same distance from the Sun. It was then that the astronomers realized that they were not dealing with planets but with fragments of planets which appeared

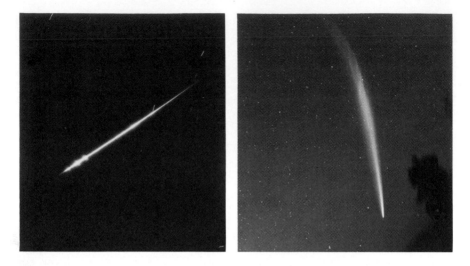

Fig. 1.1a (Left): *It was once widely believed that most bright meteors and fireballs, such as this one from the Quadrantid shower, produced meteorites. (Courtesy James Shepherd)*

Fig. 1.1b (Right): *Shower meteors are derived from the disintegration of comets, though not all cometary orbits intersect that of the Earth's. This example is Comet Ikeya-Seki 1965 VIII. (Courtesy Dr. Elizabeth Roemer)*

Fig. 1.1c: *However, closer investigation revealed that cometary debris was composed of fragile "dustballs," as shown by this model, which could not survive their plunge through the atmosphere. (Courtesy Dr. David Hughes)*

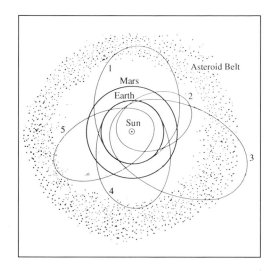

Fig. 1.1d: *Accurate orbits have since been determined for a number of meteorites which show them to be associated not with comets but with asteroids. This diagram shows the orbits of five ordinary chondrites. They are: 1) Innisfree, Alberta, Canada (fell February 5, 1977); 2) Farmington, Kansas (fell June 25, 1890); 3) Pribram, Czechoslovakia (fell April 7, 1959); 4) Lost City, Oklahoma (fell January 3, 1970—see also Figs. 2.3a–c); 5) Ohajala, Gujarat, India (fell January 28, 1976).*

Fig. 1.1e: *To an observer on Earth asteroids appear simply as star-like points of light. On this time-exposure plate asteroid 433 Eros is seen passing "close" to the star Kappa Geminorum. The long duration of the exposure reveals Eros' motion against the stellar background as a short streak. (Courtesy Harold B. Ridley)*

promising, initially at least, to those searching for a source of meteorites. Detailed work on the orbits of the asteroids, though, revealed that they could not possibly intersect the Earth and so, it was reasoned, could not provide meteorites. Researchers were forced to look elsewhere.

In 1865 Giovanni V. Schiaparelli (1835–1910) linked the Perseid meteor shower with Comet Swift-Tuttle 1862 III by comparing their orbits. It was argued that as comets passed close to the Sun, they disintegrated and left behind a stream of dust which, upon entry to the Earth's atmosphere, produce meteors. Needless to say, the meteoriticists seized upon this theory and for nearly one hundred years meteorites were assumed to be pieces of comets. The hypothesis remained largely unchallenged until the late 1950's when photographs of the fireball that preceded the meteorite fall at Pribram, Czechoslovakia, on April 7, 1959, revealed that the object had an elliptical orbit which reached out as far as the asteroid belt. Photographs of several other meteorite falls showed that they too were associated with the asteroids, and it is now generally believed that most, if not all, meteorites are asteroidal in origin.

Meteorite research continued into the twentieth century with the publication in three volumes of Emil Cohen's *Meteoriten-Kunde* (1894, 1903, 1905, Stuttgart), and Oliver C. Farrington's *Meteorites* (1915, Chicago). By this time meteoritics—a term introduced by the Russian Ju.I. Simashko in 1889—had become a recognized scientific discipline, though it earned little respect and had few practitioners, the scientific community failing to realize the importance of meteorites.

In the early 1930's two enthusiasts set out to change this attitude. Frederick C. Leonard, an astronomer at UCLA, teamed-up with Harvey H. Nininger, a former biology teacher, to form the Society for Research on Meteorites in 1933. Both men met considerable opposition from their colleagues. Leonard was warned by a senior professor not to pursue his interest in meteorites unless he wanted to ruin his career, and even the meteoritophile George P. Merrill (1854–1929), who was Curator of Geology at the Smithsonian Institution, had earlier scorned Nininger's plan to hunt meteorites with the comment, "Young man, if you live to be 100 and find one meteorite, you will have done well!" Fortunately, both Leonard and Nininger were stubborn men and were determined to investigate meteorites further. Nininger had been interested in the subject since August, 1923. After reading an article on meteorites in *Scientific Monthly* and seeing a brilliant fireball cross the night sky, his enthusiasm for meteoritics knew no bounds. Over a period of some thirty-five years, he lectured at high schools, clubs, colleges, and "to anyone else who would listen." He issued press notices and posters, established a number of useful contacts, and distributed literature to the farmers of Kansas, Nebraska, Oklahoma, Colorado, New

Mexico, and Wyoming emphasizing the importance of meteorites and offering to buy them. He estimated that he had reached between 5% and 10% of the farming community by the time his programme was completed. Nininger died on March 1, 1986, missing his centenary by less than eleven months but having recovered more than 2,000 meteorites from 226 falls. He is often seen as a popularizer of meteoritics, but in fact he undertook some important basic research and developed a number of new investigative methods. He was not always made welcome by his associates, some of whom disliked the idea of buying and selling meteorites, and after a blazing row at the 1949 Annual Meeting he left the Society which he had helped to form and could not be persuaded to rejoin until 1963.

Within five years of its founding, the Society for Research on Meteorites had attracted over 200 members, and meteoritics gained in importance, culminating in 1944 with the establishment of the Institute of Meteoritics at the University of New Mexico. Two years later the Society changed its name to the Meteoritical Society and, in 1953, published the first issue of its journal *Meteoritics*. Without the Society and the enthusiasm of its members, the science of meteoritics would undoubtedly have taken longer to evolve.

The steady growth in the popularity of meteorite collecting during the past twenty years is a reflection of the importance with which we now view meteorites. We no longer regard them simply as pieces of cosmic garbage; instead we consider them to be the hard evidence on which our current theories of Solar System evolution are based.

Meteorites are both beautiful and fascinating, and for many meteoritophiles there is something almost magical in owning a rock from space. As G.J.H. McCall noted:

> Every time we handle a meteorite, or just gaze at it, we are likely
> to feel the shiver of excitement, of the mystery of reaching out
> and making contact with the unknown.

Chapter 2

Sources of Meteorites

2.1 Introduction

Collecting meteorites is considerably easier today than it was just a few years ago. There is now a greater variety of specimens available from a larger number of suppliers, both private and commercial.

In this chapter we will look at the two principal sources of meteorites—the mineral trader and the private collector. In addition we will consider the possibility of recovering meteorites from an observed fall; discuss craters and cratering, shatter cones and impactites; and provide guidance on the legal ownership of meteorites.

2.2 Commercial Sources

A number of mineral traders throughout the world maintain a stock of meteorites, but by far the greatest concentration is in the United States. In some cases traders specialize in meteorites and advertise regularly in magazines such as *Astronomy* and *Sky & Telescope*. Most publish catalogues listing the specimens they have in stock. They also sell tektites, shatter cones, impactites, and the various accessories that are necessary for the collection, care, and investigation of specimens, including display stands, microscopes, lacquers, and other chemicals. A few traders also offer cutting, polishing, and etching services.

One of the advantages of dealing with traders is that they are often prepared to keep a record of any special requirements you may have and contact you should they feel they are able to supply your needs. Another advantage is that they will usually give advice and help on the care and preparation of specimens. Appendix C lists the names and addresses of a number of traders, but the list is by no means exhaustive.

Some traders will both sell and buy meteorites, especially if you have a specimen for sale for which they already have an interested party. Occasionally a trader will be prepared to offer exchange or part-exchange facilities, but this practice is a rarity and is confined mainly to deals between private collectors. Although selling specimens to traders is easier than finding an interested individual, you should expect to be paid less by a trader who, after all, is in business to make a profit on the deal.

Generally, traders prefer whole or near-whole meteorites. They cut the specimen into slices, each of which will be polished, sometimes etched, lacquered, and sold at a slightly higher-than-average market price. Because of this marketing practice, traders will offer various "cuts" from a meteorite (such as a slice, slab, or end piece) with or without fusion crust, some of which will be treated to reveal the internal structure. As a rule of thumb, the cost reflects the condition of the specimen, its rarity, and its weight. At the time of writing (1989), prices range from about $0.35 per gram for heavily oxidized fragments of iron meteorites to about $20 per gram for cut, polished, and lacquered slabs of pallasites. Some of the rarer meteorites, and those containing organic material, are somewhat more expensive and can reach $2000 per gram.

Table 2.1 lists the current average market prices for meteorites, but as these are subject to change, you should shop around for the best deals. Most traders offer specimens on a 10-day approval basis.

2.3   Private Sources

It goes without saying that there are far more private collectors than there are dealers. Again, the greatest concentration is in the United States, although Europe, as a whole, has quite a number. Not all collectors are interested solely in meteorites. Often they are mineral hunters who require meteorites as part of a wider collection. Once again, some collectors advertise in the popular astronomical and geological magazines, and anyone starting out in this field who wants to contact other collectors should religiously scan the small ads every month. Placing an advertisement yourself may pay dividends. A number of magazines offer free advertising to individuals, the only condition being that you must be prepared to wait until space is available before your advertisement can be published. Other magazines and popular journals make a small charge for private advertisements and are certainly worth considering.

There are three main advantages in dealing privately: First, you are likely to get a higher price for specimens which you have for sale (providing they are not above the market price, and preferably below it). Second, you are more likely to be able to exchange specimens. Third, you no longer feel

Table 2.1 Average Market Prices for Meteorites (1989)		
Type	Class	Price ($/gm)
Stones	**Chondrites**	
	Carbonaceous	2.50 – 6.00
	Olivine-Bronzite	2.50 – 9.00
	Olivine-Hypersthene	2.20 – 7.00
	Enstatite	Not Available
	Amphoterite	9.00 – 16.00
	Achondrites	Not Available
	Aubrites	6.00 – 25.00
	Diogenites	18.00 – 22.00
	Eucrites	19.00 – 40.00
	Howardites	Not Available
	Ureilites	Not Available
Stony-Irons	Mesosiderites	2.00 – 4.00
	Pallasites	1.50 – 20.00
	Lodranite	Not Available
Irons	Hexahedrites	1.00 – 5.00
	Octahedrites	0.35 – 2.25
	Ataxites	1.00 – 8.00
Misc.	Impact Glass	0.17 – 8.00
	Meteorodes	1.00 – 3.00
	Graphite Nodules	4.00 – 6.00
	Tumbled Irons	0.50 – 3.00

as though you are on your own; instead, you become part of an international fraternity of collectors who are always willing to help one another with problems. Of course, there are also disadvantages. When you first start collecting, you will not be allowed the privilege of holding onto a specimen for an approval period before making up your mind whether or not you want the meteorite. However, once you have established regular links with other collectors, a number of them will be prepared to send you specimens before you part with your hard-earned cash. It is really a matter of trust, and trust takes time to develop. Occasionally, you may find yourself paying out for a specimen before seeing it only to discover that it is not as good as you were led to believe or, worse, find that the seller does not send you the specimen and refuses to answer your correspondence. Luckily, the latter problem does not happen often, but it is a point to bear in mind. If the amount you lose is small, you should probably write it off to experience; however, you may be able to take legal action against the seller if you deem it necessary to do so. Hopefully, you will never come across this situation.

Whether you are buying or selling meteorites privately or through dealers, you should have all specimens insured against the rigors of the postal service. Meteorites should be insured against both damage and loss, and records should be kept of all transactions in a special book or computerized file, which will not only be of help should you need to make a claim but will also be of assistance when you decide to catalogue your collection. Padded envelopes should be used for small specimens, but larger meteorites should be sent in a sturdy box and wedged into place to prevent movement and damage. Newspapers are probably best for this purpose, and they have the added advantage of being inexpensive!

It is really up to the individual to decide what information should be kept on file, but as a general guide consider including the following:

For Purchases
1. Entry Number
2. Date order was placed
3. Name and address of seller
4. Details of specimen required (1st and 2nd choice)
5. Date specimen was received
6. Price paid and method of payment
7. Date paid
8. Comments

For Sales
1. Entry Number
2. Date of order
3. Name and address of buyer
4. Details of request
5. Date specimen sent
6. Price received
7. Comments

Most of the above are self explanatory. It is a good idea to encode Entry Numbers for easy identification (e.g., PAL for pallasites, C for chondrites, etc.). Comments should include details of the condition of the specimen, postal receipt numbers, etc. If you need to purchase specimens from abroad, often the cheapest method is to use a charge card such as Visa or American Express, although not all dealers use this facility.

2.4 Falls

Meteorites are classed as being either falls or finds. A fall is a meteorite whose fall has been observed. That does not necessarily mean that the

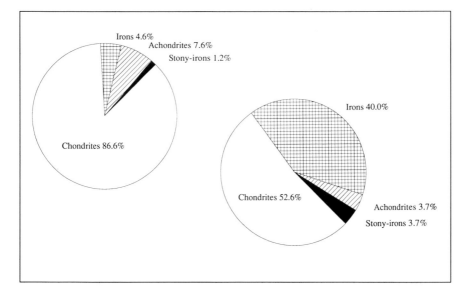

Fig. 2.1: *Relative proportions of chondrites, achondrites, irons, and stony-irons among those meteorites that were actually seen to fall (top left) and those that were found (bottom right). (Data from Catalogue of Meteorites, 4th Edition.)*

meteorite itself was seen to impact with the ground. Quite often only the preceding fireball is witnessed, the meteorite being recovered only after a careful search of the predicted impact site—something we will return to later. A find, on the other hand, is the discovery of a meteorite whose fall was not observed. Obviously, many finds are accidental, as in the case of a farmer ploughing up a buried meteorite, but in some parts of the world meteorites have accumulated in large numbers because of certain geographical features. Meteoriticists have been quick to exploit such situations.

Any report of a meteorite fall is worth investigating. Although there are twice as many finds as there are falls, the latter tend to be more representative of meteorite type (Fig. 2.1). Finds show a bias towards large, heavy, and metallic meteorites since such objects stand out more against the country rock (i.e., the local rock types and formations), which is usually silicious and not so dense. Stony meteorites obviously blend in more with the local background and hence go unnoticed. Because they are normally more friable than irons, they erode much more quickly and can decompose to little more than fragments within a few years, thus reducing the likelihood of being found. Observed falls are relatively free from this type of bias, as an observer will obviously witness a fireball event whether it results from an iron or a stony meteorite, providing he is in the right place at the right time.

Estimates as to the number of meteorite falls over the whole Earth range from about 200 to 27,500[1] per annum, depending on which data are used and how they are interpreted. Despite this number, relatively few falls are observed, however, and of these only a handful of meteorites is ever recovered. There are several reasons for this low recovery rate. Many falls occur over sparsely populated areas, and a significant proportion of all meteorites must land in the world's oceans. Cloud cover will hide the tell-tale fireballs that precede meteorite falls, and as cloud cover in the more densely populated Northern Hemisphere may gradually increase because of the Greenhouse Effect, we might expect even fewer falls to be observed in the future. Many meteorites fall during the hours of darkness when a majority of people are asleep, and lastly, people are generally unobservant—though the detonations and light show that herald the arrival of some meteorites are sufficient to at least make some people take notice!

About half of all meteorites arrive as single pieces; the others fragment in the atmosphere to form meteorite showers. Why this should be has been the subject of much debate since the end of the last century when, in 1896, E. Hauser suggested that meteorites actually enter the atmosphere as single bodies but explode because of the build-up of heat during the ablative part of their flight. The argument was based on the assumption that when a meteoroid enters the atmosphere at hypervelocity (12–72 km/sec), friction with the air is sufficient to raise the temperature of the meteoroid by several thousand degrees. Researchers had evidence of this high temperature in the form of a fusion crust which was invariably found on meteorites of all types. This "crust" is caused by minerals melting on the surface of the meteorite. While Hauser's hypothesis was an admirable attempt at solving one of the basic problems associated with meteorite falls, the evidence against thermal disruption is quite substantial. Most meteorites are reported as being only warm when collected immediately after falling, indicating that heat did not penetrate the body to any significant depth, (the exception being the Braunau iron meteorite which fell in Czechoslovakia on July 14, 1847, and was said to be too hot to touch for six hours after the fall.) This coolness may at first seem puzzling, especially in view of the fact that the surface temperature of a meteoroid during ablation can reach 4,800K, but it should be remembered that most meteorites are heated for usually less than 10 seconds at the very most. This short time period, coupled with the fact that stony meteorites are especially poor conductors of heat, explains why they remain internally cool. Even iron meteorites show little sign of a significant internal temperature rise. Laboratory tests have shown that when subjected to high temperatures, the internal structure of irons is

---

[1]The "best" estimate is probably 19,000 falls for masses in excess of 100gm: Halliday, *et al, Science* **223** 1405–07 (1984).

radically altered. The fact that this alteration is not observed in recovered iron meteorites proves that the irons do not heat through either.

In 1925, Charles P. Olivier, the Director of the American Meteor Society, proposed an alternative theory of meteoroids travelling through space in swarms, rather than as single pieces. Olivier knew that certain meteor showers display activity which suggests that the shower particles are not evenly distributed throughout the stream but tend to clump together, so that an observer watching the Perseid meteor shower, for example, would see about one meteor per minute and then perhaps three or four meteors almost simultaneously. The problem with Olivier's model was that it assumed meteorites are associated with meteor showers and, therefore, comets. However, it is now generally believed that most, if not all, meteorites are connected with the asteroids. Hence, the distribution of meteoroidal fragments scattered into a wide variety of heliocentric orbits in space is different from that found in meteoroid streams.

George P. Merrill, in 1929, was the next to theorize on meteorite showers. He proposed that some meteorites are brittle and fragment in the latter part of their flight. In 1960 the Soviet meteoriticist E.L. Krinov speculated on the existence of deep fissures tearing the meteorite apart, and more recently, in 1983, Robert Hutchison considered the possibility of meteorites splitting due to their irregular-shaped bodies being subjected to differences in the compressive forces encountered during atmospheric flight. John T. Wasson, in 1985, looked at whether collisions with other meteoroids in space could weaken their structures by causing cracks to develop within the meteoroids, and in the following year Glenn I. Huss of the American Meteorite Laboratory, Colorado, concluded that fragmentation is due to a combination of cracks, veins, fissures, and inclusions which cause the body to separate when at maximum shock loading.

The fall pattern of meteorite showers is interesting in that the fragments are not distributed randomly but are confined to a reasonably well-defined oval known as a scattering or dispersion ellipse (Fig. 2.2). While the size of the dispersion ellipse can vary considerably, the way in which the meteorites are scattered remains fairly constant. In most cases the larger, more massive meteorites will fall to the fore of the ellipse with a gradation towards smaller meteorites at the rear. This distribution appears to result from the more massive meteorites having a greater amount of kinetic energy, which carries them further down the flight path. Not all dispersion ellipses show this type of distribution, however, and as yet it is not clear whether the few variants are due to the trajectory angle with which meteorites enter the atmosphere or to other factors such as wind velocity. In the case of the Johnstown, Colorado, shower in which 27 stones fell on July 6, 1924,

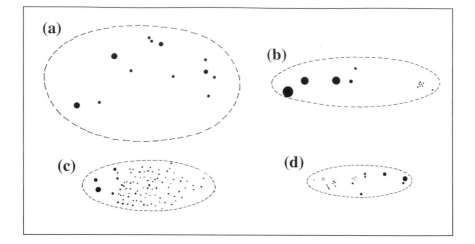

Fig. 2.2: *Dispersion ellipses for four falls: (a) Camel Donga, Western Australia (discovered January 1984), (b) Kunashak, Siberia (fell June 11, 1949), (c) Homestead, Iowa (fell February 12, 1875), and (d) Johnstown, Colorado (fell July 6, 1924). All flight directions are from right to left. Note how Johnstown is reversed. (Not to Scale)*

the distribution pattern has been reversed. Another interesting feature of dispersion ellipses is that, quite often, there is a tendency for meteorites to land in small groups within the ellipse, which is perhaps indicative of the original meteorite having undergone several stages of fragmentation.

Dispersion over an elliptical area is not confined to small meteorites, but covers the whole size range including those that cause substantial craters (e.g., Henbury, Australia). Depending on the terrain, it is not uncommon to find that the larger meteorites have produced craters with the smaller meteorites simply lying on the surface of the ground. Dispersion ellipses are virtually unknown on the airless Moon and Mercury, and where they do exist, they appear to be caused by ejecta from the main crater.

Meteorites can fragment into a considerable number of pieces. An estimated 100,000 meteorite fragments fell over an elliptical area 8 km by 1.5 km at Pultusk in Poland on January 20, 1868. In such cases the sheer number of fragments can make searches long and difficult—virtually every rock in the area has to be checked for signs of a fusion crust—but if the extent of the ellipse is known, the search can be somewhat easier.

2.5   The Recovery of Meteorites from Predicted Impact Sites

One of the first signs that a meteorite fall is about to take place is the appearance of a bright fireball. As indicators of where a meteorite will actually land, however, the fireball is notoriously deceptive. Witnesses often report that a fireball was only a few hundred meters high and "landed just over that hill." The truth is that most fireballs are at altitudes ranging from a few kilometers to several tens of kilometers, and although an object may appear to have landed quite close to an observer, it is usually the case that beyond "that hill" the fireball continued its flight for some considerable distance. Generally, then, untrained observers of fireball events are not very reliable, and any investigation of such phenomena based solely on eyewitness accounts usually contains a wide margin of error. These errors can sometimes be reduced if there are a large number of observers on either side of the fireball's path, but in the case of an exceptionally bright and impressive fireball, the errors can also compound with details of the event, often becoming subconsciously exaggerated even by experienced observers such as astronomers and meteorologists. Where there is a large time lapse between the event and interviews of the witnesses, the errors can become unmanageable. One way of overcoming this problem is to use photographs, and there are a number of camera networks in operation (such as the Florida Fireball Patrol) designed to aid the determination of fireball trajectories. Photography is by no means a foolproof method, though, as particularly bright objects lead to over-exposed images which are difficult, if not impossible, to analyze. Most networks utilize a device known as a rotating shutter on their cameras which, in effect, is a propeller that interrupts long-duration exposures several times each second. By studying the breaks which this interruption causes in the image of the fireball train, researchers can get some idea of how quickly the body decelerated and of its strength. On this basis they can decide whether or not a fireball is likely to produce a meteorite fall. At one time it was believed that most fireballs resulted in falls, but we now know that this is not the case. Meteorite-dropping fireballs are actually a comparative rarity.

When a fireball event occurs which indicates that a meteorite fall may have taken place, investigators normally send out a call for observations. If it is a moderately bright fireball, the request will be through the Scientific Event Alert Network (SEAN), the *Circulars* of The British Astronomical Association, or through a similar body to those astronomers who regularly undertake fireball and meteor observations. If the fireball was sufficiently bright to attract the attention of the general public, there will be a more widespread call for witnesses through the press and public services. (It is not unusual for the police to be swamped with calls of "flying saucers" after such events!) Investigators are mostly astronomers, meteorologists, and sometimes geologists who have had previous experience with fireball analysis. When witnesses are located, they are asked to give their exact location when they first saw the fireball and to point out the path of the object relative to certain landmarks such as trees, hills, and buildings. They are not asked for any estimates of distance or size of the object, other than relative angular size (e.g., Was the fireball as large as the full moon?), for, as noted earlier, such estimates are impossible to make accurately. Additional information as to the color, brightness, duration, audibility, fragmentation, and train features are also noted. When all the data have been collected, an attempt is made to determine the trajectory of the fireball through the atmosphere using triangulation—a relatively simple method based on trigonometric ratios—and the trajectory will be marked on a suitable map to produce a ground track. From this track the investigators will try to estimate the point of impact for the meteorite and, providing there are enough observations of sufficiently high quality, undertake to compute an orbit for the body. The chances of recovering a meteorite are improved quite significantly when there is good photographic material from two or more camera stations, though finding the meteorite is not guaranteed. A couple of contrasting examples are worthy of mention.

On the evening of January 3, 1970, a fireball was photographed by the now defunct Prairie Camera Network which was operated by the Smithsonian Astrophysical Observatory (Fig. 2.3). It seemed probable that a meteorite fall had occurred, and an impact point was calculated. A SAO researcher, Gunther Swartz, went to Lost City, the nearest town to the predicted fall area, in order to organize search parties and to ask the locals to look for anything unusual once the winter snows had melted. As he was leaving the town on January 9, he was astonished to find a 9.83 kg meteorite lying in the middle of the road only 600 meters from the predicted impact point! Eight days later a second fragment was discovered 800 meters farther away. In all, four specimens (totalling 17 kg) were recovered. The Lost City meteorites are H5 olivine-bronzite chondrites—one of the most common types of meteorite. Lost City is an exceptionally rare recovery, however, and workers must often search considerable areas in order to locate fallen objects. This is especially true where there is a delay in getting the search underway either because the raw data were difficult to obtain or because new analytical methods contradict earlier suggestions that a fireball did not survive its atmospheric plunge.

The Ridgedale event is another example. A fireball that was photographed over Canada on February 6, 1980, was not originally thought to have led to a meteorite fall, but a more detailed analysis eight years later indicated that a 1.8 kg meteorite probably landed in one piece near Ridgedale, Saskatchewan, at Latitude 53·070° N., Longitude 104·293°W. Despite intensive searches, however, the meteorite has not yet been recovered. Investigators are eager to obtain the meteorite, for its orbit was almost identical to that of the Innisfree chondrite, which fell on February 5, 1977,

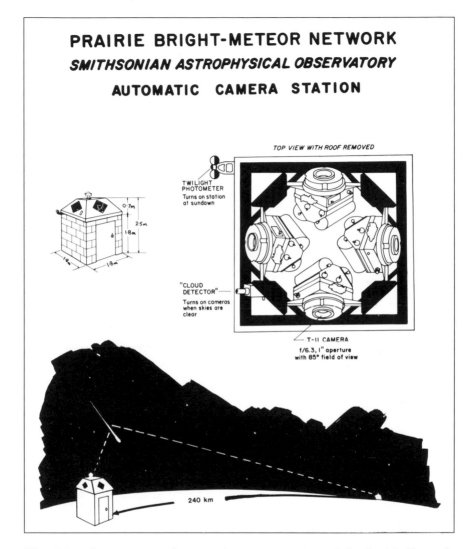

Fig. 2.3a: *Arrangement of automatic camera stations of the Prairie Network. (Courtesy Smithsonian Astrophysical Observatory)*

in Alberta, Canada. Only by a careful chemical comparison of the two meteorites can researchers decide whether they are related.

When it is suspected that a meteorite fall has taken place, it is vital to enlist help. Often the predicted impact point contains a wide margin of error which, in effect, means a large area must be searched. Professional researchers will usually turn to groups and individuals that are likely to be

Fig. 2.3b: *The Prairie Network managed to photograph this bright fireball (which heralded the Lost City meteorite fall) over Oklahoma on January 3, 1970. (Courtesy Smithsonian Astrophysical Observatory)*

interested in the nature of meteorites in order to aid the search. Interested parties may involve members of a national or local astronomical society, a geology group, staff members of museums, colleges and universities, and students with some knowledge of geology. Arguably, geology students are the most useful group, as they can usually tell when a rock looks out of place—which is normally the only indication that it is a meteorite. Search parties are asked to examine rocks for signs of ablation in the form of a fusion crust, are sometimes shown photographs and actual specimens to demonstrate what the crust looks like, and are asked to hunt for craters and sometimes damage. Metal detectors are occasionally used to locate iron

Fig. 2.3c: *Detailed trajectory analysis led to the recovery of meteorites from Lost City. (Courtesy Smithsonian Institution)*

and stony-iron meteorites. If the target area is large, a call might go out to local residents to report anything they think may be a meteorite—such as a newly found rock in an otherwise rock-free garden, though this practice does have its problems and can lead to the research team being sent large numbers of terrestrial rocks. All meteorites are photographed where they are found together with a rule or some other object (e.g., glove) to give an indication of scale. It is important not to move the meteorite until its location has been marked on a suitably large scale map and, in the case of a large number of fragments, it has been assigned a reference number. The meteorite is then put into a clean polythene bag or, preferably, wrapped in aluminum foil and sent for analysis.

2.6   Finds

A find is a meteorite whose fall has not been observed. Such objects are often discovered accidentally during commercial endeavors such as mining, prospecting, and farming (Fig. 2.4). The great meteorite hunter Harvey H. Nininger realized this fact as long ago as the 1920's and managed to secure a large number of meteorites from the agricultural community. It is not unusual to find stories and anecdotes associated with the discovery of meteorites, some of which are worth collecting almost as much as the actual specimens! One example that immediately springs to mind is the Beaver meteorite. Around 1930 a farmer in Beaver County, Oklahoma, struck a large rock while ploughing one of his fields. He continued to hit the obstacle year after year until, almost a decade later, he decided he had enough and excavated the 25.628 kg object. The local sheriff took the rock "into custody" (presumably for causing an obstruction!), and it was sentenced to life as a door stop at the county jail where it remained until 1981 when someone recognized it as a meteorite.

Although it is natural to think of finds as being simply a matter of luck, local geography can play an important rôle in the detection of meteorites. The largest meteorite ever found in the United States is the 14 tonne Willamette iron which was discovered in 1902 by Ellis Hughes in the Tualatin Valley, Oregon. Until recently it was widely believed that the meteorite was found where it originally fell. It is difficult to imagine such a massive object being moved, but some experts now take a different view. It seems at least possible that the Willamette meteorite actually landed in northern Idaho near the Canadian border. Then about 15,000 years ago, during the Ice Age, it was transported south aboard an iceberg dislodged by flood waters and eventually deposited in Oregon. Surprisingly, it would take an iceberg only 7 m square to float the massive Willamette.

An extreme example of the way in which ice can redistribute meteorites can be found in the Antarctic. Since the mid-1970's, literally thousands of meteorites have been found at the foothills of the Yamato Mountains, the Allan Hills, and Elephant Moraine. Meteorites that have fallen over the centuries have become imbedded in the ice and transported to the edge of the continent as the ice sheets advance northwards. Many meteorites are eventually deposited in the ocean, but where the ice sheet meets a mountain chain the ice is forced up-

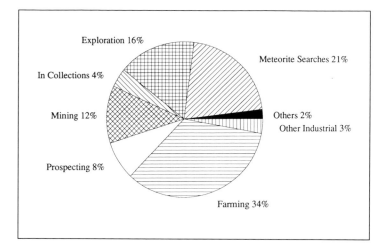

Fig. 2.4: *Relative proportions of the different kinds of* find *(non-Antarctic).*

wards and eroded, leaving the meteorites at the foothills. One of the main problems now facing researchers is "pairing"—determining which of the specimens are fragments of the same meteorite.

It is not necessary to go to the Antarctic to find large numbers of meteorites. At least 160 meteorite fragments from over 100 different falls have been discovered in parts of Roosevelt County, New Mexico, where wind has removed the topsoil. Meteoriticists are currently continuing their search of the area for new specimens.

In regions that are densely populated, finds are rare and, in some cases, totally unknown. The United Kingdom has yielded more than thirty meteorites, all of which are observed falls, and the same is true in certain parts of the United States. This situation has presumably occurred because meteorites are destroyed during industrialization and urbanization and fail to stand out against the cityscape.

These are all examples of meteorites being found in "natural" surroundings, but there are other types of finds. Every now and then, a meteorite will be "discovered" in the mineral collection of a museum or university or in a private collection either because it was originally incorrectly classified as a

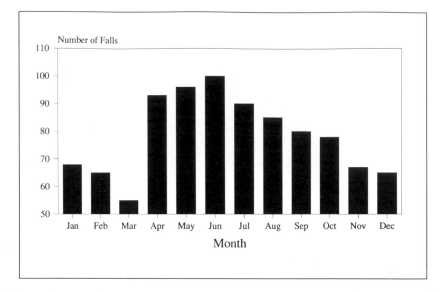

Fig. 2.5: *Monthly variation in the number of meteorite falls (AD 854–1983). The fact that most meteorites are seen to fall during the Northern summer is probably more a reflection of human outdoor activity than a true representation of the influx of meteorites.*

terrestrial rock or because it has simply been forgotten. Collectors are often presented with all manner of rocks by their neighbors in the belief that they are meteorites. In most cases the specimens will be terrestrial rocks and industrial artifacts, but there is always the possibility of being presented with a genuine meteorite which may have been used for years as a paperweight or objet d'art. The likelihood of this happening is remote, but it would be unwise not to examine anything which a neighbor finds "unusual," and any mineral collection, especially if it is very old or in the hands of an amateur, is certainly worth looking at.

2.7   Craters and Crater Fields

Many of the meteorites that strike the Earth are friable objects that cause little or no damage other than to themselves. However, some are sufficiently large and robust and impact with the ground so hard that they produce craters ranging in size from small pits to continental-size astroblemes. A complete list of authenticated and suspect craters appears in the *Catalogue of Meteorites*, but only about a dozen craters contain fragments of meteoritic matter. Many are in remote and inaccessible areas, and at least one is now the subject of a preservation order prohibiting meteorite collecting.

Fig. 2.6a: *Meteor Crater, Arizona, is the most famous and most well-preserved crater on Earth. Produced by a large meteorite plunging into the Arizona Desert perhaps about 50,000 years ago, it is 1,295m in diameter and 174m deep. (Courtesy U.S.G.S. 352)*

Cratering is an interesting and complex mechanism that is rarely mentioned in books aimed at the non-specialist. It is therefore, worth considering further here.

Most meteoroids (the term given to a meteorite before it hits the Earth) enter the atmosphere at velocities of less than 40 km/s, though it is possible for some to attain the maximum velocity of 72 km/s. Sometimes referred to as the cosmic velocity, this is the sum of the Earth's orbital velocity and that of the meteoroid. The atmosphere decelerates the meteoroid to varying degrees of efficiency, depending mainly on the cosmic velocity and the angle at which the body entered the air. During deceleration the meteoroid undergoes frictional heating as the atmospheric molecules collide with those of the meteoroid, resulting in surface temperatures of between 1600K and 4800K—sufficiently high to melt the meteoroidal minerals and thus forming a hot, viscous fluid

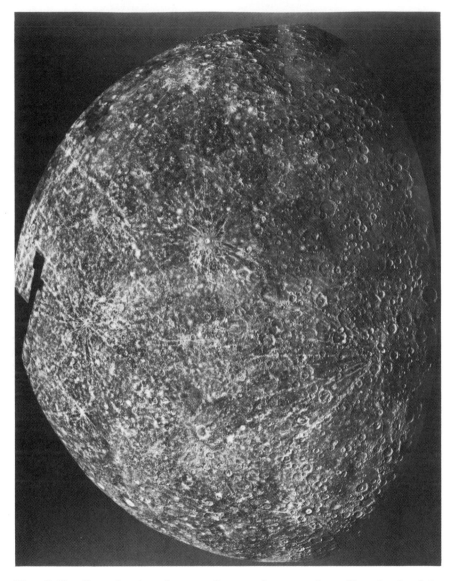

Fig. 2.6b: *Cratering is a feature of most planets and satellites in the Solar System, though the degree to which a surface is now cratered depends on whether or not the planet has an atmosphere and is geologically active. To the casual observer this may look like a photograph of the Moon but it is, in fact, Mercury.* (Courtesy NASA Jet Propulsion Laboratory)

that flows towards the rear of the meteoroid and is then released into the "smoke" train. This process is generally referred to as ablation. At these hypervelocities many of the meteoroidal and atmospheric atoms become ionized, losing their orbiting electrons and creating a long, narrow column of plasma. As the ions recombine, excess energy is released as light which observers see as a meteor or, where it is exceptionally bright, as a fireball. Eventually most meteoroids lose their cosmic velocity and fall to Earth only under the influence of the Earth's gravity. At this point the meteoroid is said to have reached its point of retardation; both ablation and ionization cease at this point and the meteor or fireball peters out. As the meteoroid falls to the ground, it rapidly cools, and the molten surface solidifies to form a fusion crust. In most cases the retardation point lies high in the atmosphere.

Although the mathematics of meteoroidal flight is beyond the scope of this book, certain generalizations can still be made. It has been shown that high velocity meteoroids are decelerated more intensely by the atmosphere than low velocity ones, but low velocity meteoroids plunge deeper into the atmosphere before reaching their retardation points. Meteoroids entering the air at oblique angles show the greatest loss of not only velocity but also mass—and mass loss also increases with increasing cosmic velocity. For small to moderate size bodies, the net result of all these actions is to reduce the cosmic velocity from several tens of kilometers per second to a terminal velocity of only a couple of hundred meters per second. The Middlesbrough meteorite, which fell in England on March 14, 1881, is a typical example. Its entry velocity was estimated to be about 34 km/s, but its terminal velocity was only 126 m/s.

Very large meteoroids never reach their retardation points but hit the ground still hot, still ablating, still ionizing, and capable of causing considerable damage. Fortunately, such events are rare.

When a meteorite hits the ground, it may produce a crater, depending not only on the impact velocity but also on the strength of the meteorite compared to that of the ground and on the nature of the target material. For chondritic (stony) meteorites with velocities of less than 100 m/s, an impact pit will be produced in soft soil comparable to the size of the meteorite. On hard ground the meteorite will fracture and may rebound at up to 80% of its impacting velocity. Increase the velocity to between 100–500 m/s and the meteorite will most probably fracture even in soft soil. Beyond about 500 m/s the diameter of the crater increases rapidly with increasing meteorite size leading to the mechanical break-up of both the soil and the projectile. Beyond about 3–4 km/s the cratering mechanism changes considerably with the crater being produced by the propagation of a shock wave through the ground. Consequently, low velocity-produced craters are called percussion craters, and high-velocity structures are known as explosion craters. This last term is poorly chosen in that it is not an explosion that causes the crater cavity but the release of pressure behind a shock front by rarefaction.

Whereas percussion craters are caused by simple compaction and redistribution of soil particles, explosion craters are far more complex. When a large meteorite makes contact with the Earth, two shock waves are produced, one of which travels into the target and the other into the meteorite. A combination of very high shock pressures (hundreds of kilobars to megabars) and free surfaces leads to a violent decompression and high velocity ejection of vaporized and molten material, giving rise to a hydrodynamic process called jetting. The jetted mass—which is composed of both terrestrial and meteoritic débris—is rapidly accelerated to a velocity in excess of the impact velocity and will still be in flight after the crater stops growing. The highest velocity material, often consisting of the smallest fragments, is excavated from the center of the crater. The area over which the débris is distributed is termed the ejecta blanket, and any observer of the Moon will know that the blanket can be of considerable size and may lead to a series of rays radiating away from the crater. Consequently, explosion craters contain only fragments of the original meteorite, but in percussion craters the entire meteorite may be found in one piece.

The most famous crater of all is, of course, Meteor Crater in Arizona (Fig. 2.6a). It was once believed that the iron meteorite that produced the crater was buried beneath the crater floor, and a small fortune was spent on attempting to find and excavate the object. What the owners of the crater did not realize was that the meteorite had vaporized on impact. The temperature rise in an impacting meteorite can be calculated using a fairly simple formula. If T is the temperature rise, v is the velocity (m/s), and C is the specific heat capacity of the meteorite (typically about 750 J/kg/K), then $T = v^2/2C$. For a meteorite with an impact velocity of only 5 km/s, (5,000 m/s) $T = 17,000$K, sufficiently high to vaporize the meteorite. Table 2.2 lists those craters in which meteoritic material has been found.

Collecting meteoritic fragments from ejecta blankets was once a profitable venture for meteoritophiles like Harvey Nininger, but today several craters are in private hands or belong to the state and are protected, having suffered considerable damage by over-enthusiastic collectors. However, meteorite ejecta are still widely available through mineral traders.

In addition to meteorites, many collectors are also interested in the by-products of cratering such as shatter cones, impactites, and high pressure/high temperature polymorphic minerals.

Located beneath crater floors, shatter cones are conical sections of rock caused by extremely high shock loading. Although they are relatively in-

Table 2.2
Craters Containing Meteoritic Fragments

Name	Discovered	Latitude	Longitude	Notes
Boxhole	1937	22° 37'S	135° 12'E	Plenty River, Northern Territory, Australia 175m diam. x 9 - 16m deep.
Dalgaranga	1923	27° 43'S	117° 15'E	Western Australia, 23m diam x 4.5m deep.
Frombork	1980	54° 20'N	19° 41'E	Vistula Lagoon, Poland 100m diam.
Haviland	1882	37° 35'N	99° 10'W	Kiowa County, Kansas, USA. Excavated site.
Henbury	1931	24° 34'S	133° 10'E	McDonnell Ranges, Northern Territory, Australia. 13 craters. Largest 201 x 110m, 12 - 15m deep.
Kaalijarv	1928	58° 24'N	22° 40'E	Saarmeaa (Oesel), Estonia, USSR.
Meteor Crater	1905	35° 03'N	111° 02'W	Coconino County, Arizona, USA. 1295m diam. x 174m deep.
Monturaqui	?	23° 57'S	68° 17'W	Atacama Desert, Antofagasta, Chile 370m diam.
Mount Darwin Crater	1933	42° 15'S	145° 36'E	Tasmania, Australia. 1 km diam. depression associated with Darwin Glass?
Odessa	1929	31° 43'N	102° 24'W	Ector County, Texas, USA. 162m diam x 6m deep.
Um-Hadid	?	21° 30'N	50° 40'E	Saudi Arabia. Less than 10m diam.
Wabar	1932	21° 30'N	50° 29'E	Rub' al Khali, Saudi Arabia. Two craters; 100m and 55m x 40m diam. Others under sand.
Wolf Creek	1947	19° 18'S	127° 46'E	Wyndham, Kimberly, Western Australia. 854m diam. x 49m deep. Iron shale found.

expensive, they do require excavating, which destroys the natural form of the crater. While it could be argued that collecting ejecta is a method of preservation in that the material would eventually erode to little more than dust, digging up shatter cones can only be seen as an act of vandalism.

Impactites were discovered in 1953 by Nininger during his investigation of Meteor Crater, Arizona. They are splash droplets, often vesicular with a superficial resemblance to volcanic bombs. They range in size from microscopic spheres to large walnuts. Grinding nearly always reveals bright nickel-iron grains set in a very fine matrix. Some impactites have a glassy composition.

Meteorodes are an extremely rare form of débris. In the case of the Haviland, Kansas crater they consist of broken, fissured, and complete crystals of olivine set in a light-colored porous oxide surrounded by a mammilated oxide shell. They are associated with the Brenham pallasite.

Minerals that have been produced by shock loading, such as stishovite and coesite, are available from some mineral dealers. They have been used by researchers as proof that some craters are meteoric in origin.

2.8 Legal Ownership

If a meteorite lands in someone's backyard, to whom does it belong? The law dealing with this situation differs considerably from one country to the next.

Court cases in the United States have established that the owner is the person on whose land the meteorite was found, and as most of the land in the western U.S. is still United States Government property, meteorites discovered there are vested in the Smithsonian Institution. In a number of other countries, meteorites belong to the state whether or not they land on private property, and some states encourage finders to hand in meteorites by offering a reward. In Britain there is no legislation, a situation which can cause considerable problems. For example, a property owner may not necessarily own the mineral rights of the ground on which the property stands. As a meteorite is technically a "mineral," it could be argued that the property owner has no claim to the object. In most cases it probably does not matter, as meteorites have such relatively little value that litigation would be unacceptably expensive. However, as small fortunes have been spent on proving points of principle, mineral rights issues are worth bearing in mind. Whatever the legal situation, any genuine meteoritophile is under a moral obligation to make all new finds and falls available for research. In most cases professional meteoriticists and institutions are willing to prepare, cut, and polish sections for donors rather than miss the opportunity of acquiring a new specimen.

Chapter 3

Composition

3.1 Introduction

Meteorites are mineral aggregates that show considerable diversity. Although the layman tends to think of minerals as rock-like compounds of various elements, in the complex world of mineralogy certain "native" elements can themselves be classed as "minerals." Copper is a typical example, as is carbon, both of which can be found in meteorites. For the most part meteorites are basically similar to terrestrial rocks, but in detail there are a number of important differences. As a result we find that certain common terrestrial minerals, such as quartz, are relatively rare in meteorites, and, similarly, meteorites contain a number of minerals that are totally alien to the surface of our planet. It is perhaps reassuring to find, however, that those mineral groups that are important to the rock-building processes of the Earth, such as the pyroxenes and olivines, are equally important to meteorites. Tables 3.1 and 3.2 show typical chemical compositions for iron and stony meteorites.

3.2 Isomorphism and Polymorphism

Although mineralogical theory is beyond the scope of this book, it is necessary to say something of two important properties of minerals: isomorphism and polymorphism.

Isomorphism is a property of minerals in which two minerals can have different chemical compositions but essentially the same structure. The olivines are said to be isomorphous. Their formulae change from pure Mg_2SiO_4 (commonly called forsterite) to pure Fe_2SiO_4 (fayalite). This situation occurs because the iron and magnesium ions are very similar in size ($Fe^{2+} = 74$ pm, $Mg^{2+} = 66$ pm; where 1 pm $= 10^{-12}$m) and can

| Table 3.1 | |
| Typical Iron Meteorite | |
Element Composition	
Element	Percent
Fe	89.50
Ni	9.32
Co	0.66
Cu	0.04
Others	0.48

| Table 3.2 | | |
Typical Element Composition of Chondritic (stony) Meteorites Compared to Igneous Rock Composition		
Element	Chondrites (%)	Igneous (%)
O	34.55	46.60
Fe	26.86	5.00
Si	17.62	27.74
Mg	14.02	2.06
S	2.06	0.06
Ca	1.20	3.63
Ni	1.37	0.08
Al	1.21	8.16
Na	0.57	2.80
Cr	0.25	0.02
C	0.1	0.03
Others	0.19	3.82

therefore substitute for one another within the atomic structure of the olivines. Where complete ionic substitution can occur between two "end-members"—in this case forsterite and fayalite—a range of mixed crystals may form, producing what is referred to as a solid solution series. For the olivines the variability in composition is expressed as $(Mg, Fe)_2SiO_4$. As can be seen from Table 3.3, there are six different "varieties" of olivine, depending on the relative proportions of iron and magnesium, but by convention the composition of an olivine is indicated by specifying its fayalite content. Thus, an olivine with a molecular fayalite content of 15% (Fa_{15}) is, in fact, the magnesium-rich chrysolite variety.

Polymorphism is a property of minerals in which a single element or compound may occur in different crystal forms because of different packing arrangements of the constituent atoms, ions, and molecules. These structural differences are usually dependent on temperature and pressure, with

Table 3.3 — Part A
Minerals Found in Meteorites

Group	Mineral	Abb.	Formula	Notes
Elements	Carbon		C	
	Graphite		C	Low-pressure polymorph
	Cliftonite		C	Low-pressure polymorph
	Diamond		C	High-pressure polymorph
	Lonsdaleite		C	High-pressure polymorph
	Copper		Cu	Isolted grains
	Nickel-iron			
	Kamacite		$Ni_{0.07-0.04}Fe_{0.93-0.96}$	
	Taenite		$Ni_{0.2}Fe_{0.8}$	
	Sulfur		S	
Oxides	Magnetite[1]		Fe_3O_4	Mainly present in fusion crust
	Chromite[1]		$FeCr_2O_4$	
	Spinel[1]		$MgAl_2O_4$	
	Ilmenite		$FeTiO_3$	
	Rutile		TiO_2	
Silicates	Quartz		SiO_2	Low-pressure polymorp
	Stishovite		SiO_2	High-pressure polymorph
	Coesite		SiO_2	Very high-pressure polymorph
	Tridymite		SiO_2	High-temperature polymorph
	Cristobalite		SiO_2	Very high-temperature polymorph
	Feldspar			
	Plagioclase		$NaAlSi_3O_8 - CaAl_2Si_2O_8$	Solid-solution series
	(Albite)	Ab	$NaAlSi_3O_8$	Na-plagioclase end-member
	Anorthite	An	$CaAl_2Si_2O_8$	Ca-plagioclase end-member
	(Albite		$Ab_{100-90}An_{0-10}$)	Unknown in meteorites
	Oligoclase		$Ab_{70-90}An_{10-30}$	
	Andesine		$Ab_{50-70}An_{30-50}$	
	Labradorite		$Ab_{30-50}An_{50-70}$	
	Bytownite		$Ab_{10-30}An_{70-90}$	
	Anorthite		$Ab_{0-10}An_{90-100}$	Rare

Table 3.3 — Part B
Minerals Found in Meteorites

Group	Mineral	Abb.	Formula	Notes
Silicates	*Maskelynite		$NaAlSi_3O_8 - CaAlSi_3O_8$	Glass of plagioclase composition. Not a true mineral
(cont.)	Olivines			Solid-solution series
	Forsterite	Fo	Mg_2SiO_4	Magnesium end-member
	Fayalite	Fa	Fe_2SiO_4	Iron end-member
	Forsterite[2]		$Fa_{0-10}Fo_{100-90}$	
	Chrysolite[2]		$Fa_{10-30}Fo_{90-70}$	
	Hyalosiderite[2]		$Fa_{30-50}Fo_{70-50}$	
	Hortonolite[2]		$Fa_{50-70}Fo_{50-30}$	
	Ferrohortonolite[2]		$Fa_{70-90}Fo_{30-10}$	
	Fayalite		$Fa_{90-100}FO_{10-0}$	
	Pyroxenes	px	$(Mg, Fe)SiO_3$	
	Orthopyroxenes	opx		$Ca^{2+} < 5\%$ Orthorhombic crystals
	Enstatite	En	$MgSiO_3$	FeO content
	Hypersthene		$(Mg, Fe)SiO_3$	varies between
	Bronzite		$(Mg, Fe)SiO_3$	0–25%
	Ferrosilite	Fs	$FeSiO_3$	approx
	Clinopyroxenes	cpx		Monoclinic crystal system
	Clinoenstatite		$MgSiO_3$	$Ca^{2+} < 3\%$
	Clinohypersthene		$(Mg, Fe)SiO_3$	
	Clinobronzite		$(Mg, Fe)SiO_3$	
	Clinoferrosilite		$FeSiO_3$	
	Augite		$(Ca, Mg, Fe)SiO_3$	Ca-rich with some aluminum
	Diopside		$(Ca, Mg)SiO_3$	Ca^{2+} substitution 15–50% (Fe + Mg)
	Pigeonite		$(Ca, Mg, Fe)SiO_3$	Ca-poor
	Hedenbergite		$FeCa(SiO_3)_2$	Ca/(Mg + Fe)
	Wollastonite	Wo	$CaSiO_3$	5–15/(95–85)
	"Serpentines"		$Mg_6Si_4O_{10}(OH)_8$	
	Gehlenite		$MgSi_2O_7$	
	Sodalite		$Na_8Al_6Si_6O_{24}Cl_2$	
	Nepheline		$NaAlSiO_4$	
	Garnet		$Ca_3Al_2Si_3O_{12}$	Grossular
	Perovskite		$CaTiO_3$	

Table 3.3 — Part C
Minerals Found in Meteorites

Group	Mineral	Abb.	Formula	Notes
Sillicide	Perryite		$[Fe(Ni)]_2Si$	
Nitride	Osbornite		TiN	
Sulfides	*Oldhamite		CaS	Oxidizes to gypsum
	Pyrrhotite		FeS	
	Troilite		FeS	
	*Daubréelite		$FeCr_2S_4$	
	Pentlandite		$(Fe, Ni)_9S_8$	
	Alabandite		MnS	
Phosphides	*Schreibersite		$(Fe, Ni, Co)_3P$	Needle-like crystals are called rhabdite
Phosphates	*Merrillite		$Na_2Ca_3(PO_4)_2$	
	*Farringtonite		$Mg_3(PO_4)_2$	
	Chlorapatite		$Ca_5(PO_4)_3Cl$	
Carbides	Cohenite		Fe_3C	
	*Moissanite		SiC	
Carbonates	Dolomite		$CaMg(CO_3)_2$	
	Magnesite		$MgCo_3$	
	Calcite		$CaCO_3$	
Chloride	*Lawrencite		$FeCl_2$	Decomposes to ferric chloride and limonite. Buchwald believes it to be a product of terrestrial alterations.
Hydrous Minerals	Gypsum		$CaSO_4 \cdot 2H_2O$	
	Epsomite		$MgSO_4 \cdot 7H_2O$	
	Bloedite		$Na_2Mg(SO_4)_2 \cdot 4H_2O$	

* These are not found on Earth, named after: Daubréelite, Gabriel-Auguste Daubrée (1814–1896); Farringtonite, Oliver Cummings Farrington (1864–1933); Lawrencite, John Lawrence Smith (1818–1883); Maskelynite, M.H. Nevil Story-Maskelyne (1823–1911); Merrillite, George Perkins Merrill (1854–1929); Moissanite, H. Moissan; Rhabdite; Greek (rhabdos = rod); Schreibersite, Carl Franz Anton von Schreibers (1775–1852).

1 Structurally isomorphous "spinel group."

2 These terms are only rarely used in modern texts.

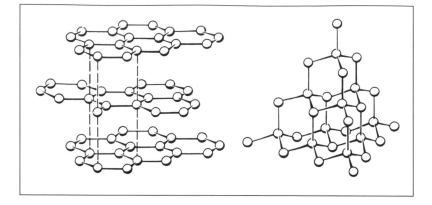

Fig. 3.1: *Polymorphs of carbon. Shown left is the low-density, weakly-bonded graphite. Right is the hight-density, strongly-bonded diamond.*

low temperatures and high pressures producing the most compact, high density structures. Carbon, for example, can take the form of graphite (low-pressure) and diamond (high-pressure) (Fig. 3.1).

When a meteorite is subjected to shock loading, some of the minerals can undergo a transformation. Graphite can change to diamond, and certain other minerals can produce glass-like substances. The effects of some of these changes are given in Table 3.4.

Table 3.4	
Shock-induced Changes in Meteorites	
Minimum Shock Pressure (kb)	Shock Indicators
-	Blackening of chondrites
-	Orthopyroxene converts to clinopyroxene
-	Melted metal/troilite (Could also be caused by non-shock heating)
130	Graphite converts to cubic diamond
135	Pyroxene converts to majorite
300	Plagioclase converts to maskelynite
450	Olivine converts to ringwoodite
450	Plagioclase converts to glass
450	Pyroxene converts to glass
700	Graphite converts to hexagonal diamond

3.3 Minerals

Carbon is present in several classes of meteorites though usually in small amounts. Most chondrites (stones) have a bulk carbon content of less than

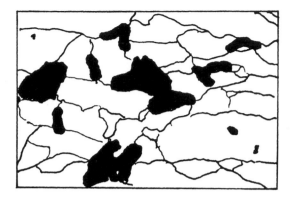

Fig. 3.2: *Sketch of the internal structure of a ureilite showing orientation of the crystals and the presence of carbon.*

2%, with the exception of Type 1 carbonaceous chondrites, which have 3–5%. Despite the small amount, carbon is largely responsible for the blackening of chondritic meteorites. The low-pressure polymorph graphite is commonly found as an accessory mineral in irons. It can occur as plates or as minute pseudocubic crystals termed cliftonite, but more often it takes the form of nodules. Usually these are smaller than a pea, but larger nodules can be found, the record being a 92 gm nodule from the Cosby's Creek octahedrite. Dissolving an iron in acid will sometimes leave behind flakes of carbon.

Microscopic crystals of diamond are also found in meteorites. They were first discovered in 1888 by N. Jerofejeff and P.A. Latchinoff while examining the Novo-Urei stone. The ureilites—as this type of meteorite became known—also contain graphite (Fig. 3.2) and the ultra-high-pressure carbon polymorph lonsdaleite. Researchers have argued that the presence of diamond and lonsdaleite indicates intense shock pressures probably encountered when the meteorites' parent bodies disintegrated. In meteoritic fragments recovered from explosion craters, these polymorphs could also be produced by the shock of impact.

Most native metals (i.e., metals occurring in a free state) are relatively rare. The presence of small isolated grains of copper was, for many years, erroneously considered to be foreign to meteorites, introduced when specimens were cut using copper saws. On polished slabs, however, the grains of copper show well against darker matrix material. Other metals, such as gold, are present only in trace amounts.

Reference is often made to "iron" meteorites, but in reality they are composed of nickel-iron alloys. The α-phase is the Ni-poor mineral kamacite (Ni $\sim$6%), while the γ-phase is called taenite (Ni = 13–48%). These phases are evident in the form of the Widmanstätten pattern (more on this in

Chapter 4).

Of the oxides the most abundant is magnetite, which is important to Type 1 carbonaceous chondrites. In most other meteorites, it is found only in the fusion crust. The variety of chromite in stony meteorites is usually rich in aluminum and magnesium but, again, is present only in small amounts.

The silicates represent one of the most important mineral groups, but their occurrence in meteorites is somewhat different from that normally found in terrestrial rocks. Quartz, so common in the Earth's surface material, has been found only in enstatite chondrites, as has cristobalite. Both minerals are polymorphs of silica, SiO_2, with quartz forming at low temperatures. Other polymorphs are known: the high-temperature tridymite is present in several types of meteorites, while stishovite and coesite are formed at high pressures and can sometimes be found within explosion craters (see Chapter 2.7).

The feldspars comprise another important group (Fig. 3.3). These are aluminum silicates containing variable amounts of sodium and calcium, thus forming an isomorphous solid-solution series from albite ($NaAlSi_3O_8$) to anorthite ($CaAl_2Si_2O_8$). A third end-member, the potassium-rich orthoclase ($KAlSi_3O_8$), is exceedingly rare in meteorites. Albite is not found in meteorites, and both anorthite and labradorite are rare. The most common feldspar in the chondritic meteorites is oligoclase; in the howardites, eucrites and mesosiderites, bytownite is most common. The nakhlites contain andesine. Plagioclase that has been subjected to severe shock transforms into a translucent glass called maskelynite, which is unknown on Earth. It appears isotropic when viewed in polarized light.

A second isomorphous series is that of the olivines (Fig. 3.4)—the most abundant silicate in meteorites. As mentioned above, the end-members of this series are the magnesium-rich forsterite and the iron-rich fayalite. The former is found mainly in ureilites, mesosiderites and rarely in enstatite achondrites. Hyalosiderite and hortonolite are common in amphoterites and Type 3 carbonaceous chondrites, with chrysolite occurring mainly in ordinary chondrites and pallasites.

The pyroxenes, another important group, occur in two principal forms: orthopyroxene, which has an orthorhombic crystal system and a Ca^{2+} content of less than 5%, and clinopyroxene with a monoclinic crystal system which contains almost the maximum *limit* of 50% replacement of Fe^{2+} and Mg^{2+} by Ca^{2+}. The orthopyroxenes enstatite, hyperstene, and bronzite are found in considerable amounts in some meteorites, enstatite having the lowest Fe content and hypersthene the highest. Closely related to these minerals are the clinopyroxenes augite (Ca-rich), pigeonite (Ca-poor), and diopside (Fe-free). Of the sulfides, oldhamite and daubréelite

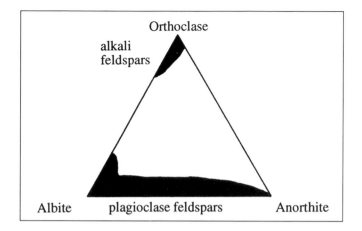

Fig. 3.3: *Triangular diagram of feldspar composition. Feldspars at the bottom left-hand corner are pure albite (Na-rich). Through ionic substitution of Ca for Na, the composition changes to pure anorthite (Ca-rich), the composition range being shown in black. Also shown are the alkali feldspars (K-rich), which are rarely present in meteorites.*

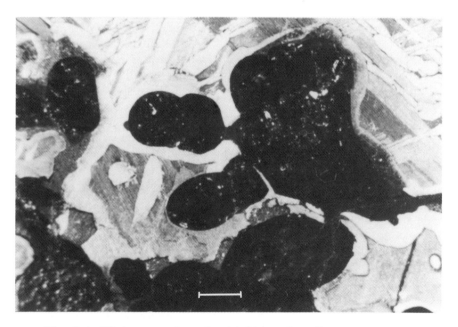

Fig. 3.4: *Olivine crystals set in nickel-iron of a pallasite. Bar = 1mm*

do not occur naturally on Earth. Oldhamite, discovered in 1870 by Maske-lyne in the Bustee achondrite, is present in enstatite chondrites and en-statite achondrites. It appears light brown and translucent when fresh, and in contact with water it produces gypsum.

Daubréelite was discovered in 1876 by J. Lawrence Smith in the Coahuila meteorite. It occurs mainly in highly-reduced meteoritic as-semblages, especially in hexahedritic irons, and is often associated with troilite as parallel intergrowths, or adjoining it and sometimes bordering it. Daubréelite will dissolve in nitric acid but not in hydrochloric acid, though professional researchers prefer to use non-destructive methods such as the microprobe. The oxide chromite, which is often Al- and Mg-rich, oc-curs in oxidized stones because of ionic substitution. Note that daubréelite ($FeCr_2S_4$) is the sulfide analogue of chromite ($FeCr_2O_4$).

Troilite is an important sulfide which is similar to terrestrial pyrrhotite. It is named after its discoverer, Father Domenico Troili, who first observed it in the Albareto meteorite. Troilite is present as an accessory mineral in most meteorites and can occur as plates, small irregular grains in chondrites, and nodules up to about 10 cm diameter in irons. When fresh it has an unmistakable bronze-colored luster.

The phosphide schreibersite is another non-terrestrial mineral which, although present in enstatite chondrites and enstatite achondrites, is more commonly found in irons, especially hexahedrites. It was first detected in the Magura iron by W. Haidinger in 1847 and can occur in a wide variety of shapes including shells enveloping troilite nodules, independent radiat-ing intergrowths, very thin platelets, and minute needle-like, carbon-rich crystals called rhabdite, which are oriented in sympathy with the Wid-manstätten structure of the octahedrites. This orientation suggests sepa-ration from the iron after solidification—a process known as "exsolution." Schreibersite is easily recognized in irons. It is hard and brittle, breaks quite easily, and is insoluble in cold dilute acids. When fresh it has a bright silvery-white luster which tarnishes to a yellowish-bronze.

Often mistaken for schreibersite is the carbide cohenite, the difference being that cohenite is soluble in copper chloride-ammonium chloride. It is as rare in meteorites as it is in the Earth and occurs mainly in Ni-poor irons. The silicon carbide moissanite was discovered in 1904 by H. Moissan. It occurs as small hexagonal crystals which range from pale green to emerald green in color.

The phosphorous content of meteorites is uniformly low, as reflected in the fact that the few phosphates that do exist are found only in accessory amounts. The colorless merrillite was first recorded by G. Tschermak in 1883 but was named by E.T. Wherry in 1917 in honor of George P. Merrill, the Curator of Geology at the Smithsonian Institution. It is visually and optically very similar to chlorapatite. Farringtonite was discovered only as recently as 1960 in the Springwater pallasite by E.R. DuFresne and S.K. Roy.

Yet another terrestrially unknown mineral is lawrencite which, because it readily absorbs moisture from the air, is usually found as small greenish droplets. It poses a problem for collectors in that it causes iron meteorites to rust and dissolve at an alarming rate. When exposed to moisture the ferrous chloride converts to ferric chloride and limonite, an iron hydroxide. Further reactions cause the ferric chloride to produce more limonite and hydrochloric acid, which rapidly erodes the specimen. The preservation of meteorites containing lawrencite will be discussed in Chapter 6.

The hydrated sulphates epsomite and bloedite are found only in Type 1 carbonaceous chondrites, while gypsum occurs in enstatite chondrites and enstatite achondrites as a result of the oxidation of oldhamite.

3.4 Isotopes and Meteorite Ages

The age of a meteorite may be determined by applying standard geological radiometric dating techniques. Although such experiments are beyond the ability of the amateur, it is perhaps worth reviewing the methods here, as most accounts aimed at the non-specialist tend to be rather vague.

Minerals containing radioactive isotopes have, in effect, a built-in "clock" that can be used to determine when the mineral formed. The principle is very simple. Isotopes decay at a constant exponential rate known as the half-life. The half-lives of all the common radioactive isotopes have been determined by laboratory experiments on pure samples. If the original abundance of a radioactive isotope in a specimen is known, then the age of the specimen can be calculated from its current radioactivity using the known decay rate for that isotope. One such decay mode is:

$$^{235}U \longrightarrow \text{decays to} \longrightarrow ^{207}Pb$$

In this mode uranium-235 (the parent) decays to produce lead-207 (the daughter). The time taken for exactly one-half of the original uranium to decay to lead is 704 Ma. Other commonly used isotopes are given in Table 3.5.

The abundance of an isotope can be determined by using a mass spectrometer, of which there are several different kinds. In a time-of-flight spectrometer (Fig. 3.5) the atoms of the isotope are ionized and projected down an evacuated drift tube. Inside the tube the ions separate according to mass with the more massive ions lagging behind the lighter ones. At the end of

Table 3.5 Isotopes Commonly Used in Radiometric Dating		
Parent	Daughter	Half-life (Ma)
^{14}C	^{14}N	5.73×10^{-3}
^{26}Al	^{26}Mg	7.40×10^{-1}
^{129}I	^{129}Xe	17
^{40}K	^{40}Ar $+^{40}$Ca	1,250
^{87}Rb	^{87}Sr	49,000
^{232}Th	^{208}Pb $+ 6^4$He	13,900
^{235}U	^{207}Pb $+ 7^4$He	704
^{238}U	^{206}Pb $+ 8^4$He	4,510

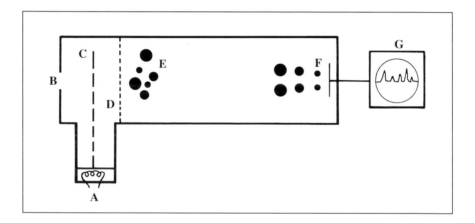

Fig. 3.5: *A time-of-flight mass spectrometer. Filament* **A** *produces electrons which are dispatched towards the gas introduced at* **B**. *The gas is ionized,* **C**, *and accelerated by plate* **D**. *Although the ions are mixed at first,* **E**, *by the time they have travelled down the tube and reached the target,* **F**, *they have separated according to mass with the more massive ions lagging behind. A monitor,* **G**, *gives a read-out of the abundance of each group of ions.*

the drift tube is a receptor which counts the number of ions and estimates their flight time (velocity) and then uses this information to determine the mass of the ions and their abundance. In this way the parent:daughter ratio can be found, which can then be used to calculate the age of the meteorite using the following equation:

$$t = \tau \, \frac{\log(p/d)}{\log(1/2)}$$

where t = Age of the meteorite, τ = Half-life, p = Parent isotope abundance, and d = daughter isotope abundance.

There are, however, various problems involved in estimating ages. For example, some ^{207}Pb may already have been present in the sample at the time of its formation. Furthermore, geologic processes that involve heat tend to re-set the isotopic clocks. Consider those isotopes that produce gaseous daughter nuclides. The gases become trapped within the crystal lattice structures of minerals and diffuse out very slowly. However, since lattice diffusion is highly temperature-dependent, the abundance of gaseous daughters is more a record of the thermal history of a meteorite and not necessarily an indication of when the meteorite formed. Consequently, researchers need to exercise care in choosing the correct isotopes and, where possible, employ several different dating techniques. Nearly all meteorites appear to be about 4,600 Ma old, which is in agreement with current theories on the origin of the Earth and Solar System, but some meteorites—the SNC suite (shergottites, nakhlites, and Chassigny)—appear to be only about 1,300 Ma old. A careful examination of the SNC meteorites suggests that they came from a planet that was geologically active until fairly recently. As gases trapped within these meteorites are found in the same concentrations as in the atmosphere of Mars, it looks as though they may have come from the Red Planet. More will be said on this subject in Chapter 5.

In 1973, Robert Clayton and his colleagues at the University of Chicago discovered anomalous ^{16}O in the chondrules and Ca- and Al-rich inclusions of certain carbonaceous chondrites. Since then other anomalous isotopes have been found which could not have formed in sufficient quantities in the Solar System. This discovery has led some researchers to suggest that the isotopes were "seeded" into the Solar System prior to its formation possibly as débris from the explosion of a nearby star (a supernova). The shock wave from the explosion would ultimately have led to the contraction of the dusty and gaseous nebula, which eventually gave birth to the sun and planets.

3.4.1 Cosmic Rays

The isotopes mentioned above are useful in determining when meteorites formed. However, a second group of isotopes can reveal when the parent bodies of the meteorites disintegrated and how long meteorites existed in space as meter-size fragments.

Cosmic rays are mostly very high energy protons which come from the sun and the rest of the galaxy. They are absorbed by meteoroids while in space, but those meteoroidal atoms that are struck by the rays undergo a se-

Table 3.6 Cosmic Ray Exposure Ages of Various Classes of Meteorite	
Meteorite Class	Age Clusters (Ma)[1]
Irons	500–590
Irons	240–290
Medium Octahedrites	≥ 300
Hexahedrites	≤ 300
Ataxites	≤ 300
Stones	20–30
Chondrites	22
L-Chondrites	≤ 50
Eucrites	≤ 50

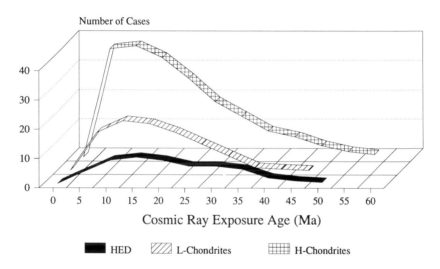

Fig. 3.6: *Cosmic Ray Exposure Ages of three different types of meteorite. The large peak for the H-Chondrites indicates a single catastrophic event whereas the other curves suggest that L-Chondrites and HED (howardites, eucrites, and diogenites) have undergone a series of destructive episodes.*

ries of changes producing a range of cosmogenic nuclides—a process known as spallation. Since the rays will only penetrate up to about one meter of rock, any material buried deeper than this is shielded from the protons. Consequently, the abundance of cosmogenic nuclides serves as a record of

---

[1]Ma = Million years

how long a meteoroid has been exposed to the cosmic environment. Table 3.6 gives the cosmic ray exposure ages of several different types of meteorite and shows, perhaps not surprisingly, that iron meteorites are more robust than stones and have gone through fewer stages of fragmentation. Fig 3.6 shows the exposure ages for several classes of stony meteorite. The large, well-defined peak of the H-chondrites indicates that they underwent only one major disruptive event, whereas the other groups went through several.

The type and abundance of cosmogenic nuclides can also be used to estimate how long a meteorite has been on the Earth's surface, as the atmosphere protects meteorites from cosmic rays and therefore prevents further spallation.

Cosmic rays also damage mineral crystals, leaving behind tell-tale tracks of their presence, the extent of which can give some indication of the pre-atmospheric size of a meteorite. H. Bhandari and colleagues recently examined the Gujargaon chondrite for particle tracks and found that, prior to atmospheric entry, the body was about 18-20 cm in diameter but lost 3.1 cm on the leading surface and 1.2 cm on the rear surface during ablation.

3.5 Do Meteorites Smell? — An Aside

Few modern reports of meteorite falls make any mention of odor, but in the past meteorites were quite often said to smell of sulfur. This description could be due to the result of the imagination of witnesses who thought meteorites were the "Devil's work" and therefore carried the stench of Hell, or it could be a real effect caused by fused troilite (FeS), but, for some unexplained reason, investigators have failed to record the smell in recent years. Obviously more research is needed!

Chapter 4

Structure and Appearance

4.1 Introduction

Many meteorites display structures that have no terrestrial equivalent. Externally their features are sculptured by their brief hypervelocity flight through the atmosphere, but internally they have been shaped by exposure to pressures, temperatures, and catastrophic events totally alien to the rocks of the Earth's surface. These distinctive features can often be used to positively identify meteorites without having to resort to complex chemical analysis.

4.2 External Features

The high temperatures and pressures experienced by a meteoroid in flight are sufficient to change completely its external appearance. The heat generated during ablation (1600–4800K) melts the minerals near the meteoroid's surface, and these are then swept backwards by the air flow to be deposited in the so-called "smoke train." When the meteoroid reaches its retardation point (Chapter 2.7), ablation ceases, and the molten minerals fuse together, forming a thin skin over the entire meteoroid.

4.2.1 Fusion Crust (Fig. 4.1)

Meteorites that have been recovered soon after their fall are nearly always enveloped in a crust of fused minerals. The crust is invariably thin, dull, or lustrous, and can cover the entire surface. In some cases, however, the crust may have spalled away because of thermal fatigue or fragmentation of the body. If this happens while the meteoroid is still ablating, a second, thinner, crust can develop on the freshly exposed surfaces which is usually lighter in color than the primary crust. Immediately below the fusion crust is a zone less than 1 cm deep that has been thermally altered.

Fig. 4.1a–c: *Views of some of the 75 eucrites which fell at Pasamonte, New Mexico on March 24, 1933, showing porous, glassy, fusion crusts (warty) with numerous gas holes. (Courtesy Smithsonian Institution, Museum of Natural History)*

Fig. 4.1d: *One of over 300 meteorites which fell at Nuevo Mercurio, Mexico, on December 15, 1978. This H5 olivine-bronzite chondrite shows a ribbed fusion crust. See Fig. 4.4a for contrast between fusion crust and interior. (Courtesy Smithsonian Institution, Museum of Natural History)*

In stones the crust consists of a glass-like substance of broadly the same composition as the meteorite, which is normally less than 1 mm thick but which can, on occasions, be up to 10 mm, depending on the stone's ability to conduct heat. In newly-fallen irons the crust is composed of a magnetic iron oxide, Fe_3O_4, which quickly deteriorates, and is replaced by rust.

Fusion crusts are more notable on stones, contrasting with the lighter interior, except for aubrites in which the crust is white or transparent because of the lack of iron. In oriented meteorites (see 4.2.3 below) the crust is thinner at the nose and thickens towards the rear as material that is swept backwards solidifies during non-ablative flight. The color of the crust can give some indication of the composition of the meteorite (Table 4.1).

Table 4.1	
Fusion Crusts and Meteorite Class	
Color	Class
Black and glossy	Howardites, eucrites, nakhlites
Matt black	Chondrites
Black to bluish	Irons
White or transparent	Aubrites

Fig. 4.2a: *Meteorites that have been exposed to the environment lose their charac-teristic fusion crust. This photograph of the Bayard L5 chondrite shows prominent cracks and crustal flaking, both of which are evidence of weathering. (Courtesy Dr. A.W. Struempler, Chadron State College)*

Fig. 4.2b: *Oxidized fragment of the Odessa, Texas, iron meteorite.*

Long-term exposure to the Earth's environment invariably destroys fusion crusts (Figs. 4.2a, b). If the meteorite is buried within the Earth, circulating groundwater converts the crust into a white caliche encrustation. In irons a shale develops which easily flakes and may form a thick shale bed beneath the meteorite. When exposed to air, irons readily oxidize and turn brown, but in arid climates fresh metal can usually be found underneath a very thin oxidized film. The weathering of Antarctic meteorites is classed according to the amount of rust visible to the naked eye, with "A", "B", and "C" representing "minor," "moderate," and "severe" rustiness. In about 4% of Antarctic meteorites, white, powdery, evaporitic deposits can be found on exposed surfaces, the compositions of which are given in Table 4.2

Table 4.2 Evaporite Minerals Identified in Antarctic Meteorites	
Mineral	Chemical Formula
Epsomite	$MgSO_4 \cdot 7H_2O$
Gypsum	$CaSO_4 \cdot 2H_2O$
Hydromagnesite	$Mg_5(CO_3)_4(OH)_2 \cdot 4H_2O$
Nesquehonite	$Mg(HCO_3)(OH) \cdot 2H_2O$
Starkeyite	$MgSO_4 \cdot 4H_2O$

Terrestrial weathering of meteorites is a problem not only for collectors but also for researchers. Weathering redistributes chemical elements and transforms original minerals to alteration products, which hinder the investigation of specimens.

The Russian meteoriticist E.L. Krinov developed a classification system for fusion crusts as follows:

I. FRONTAL SURFACES

    A. Close Textured—Smooth crust showing no relief but which is sometimes marked by colored streaks that radiate from the nose. This type of crust is characteristic of irons and may be caused by high atmospheric pressures.

    B. Nodular or Knobby—Mainly found on stones, an otherwise smooth surface is disrupted by fine angular nodules which coincide with metal concentrations. Again, this may be due to high atmospheric pressures.

II. LATERAL SURFACES

    A. Striated—Ridge and groove striations on a close textured crust mark the air flow across the lateral surfaces of some meteorites. The striations tend to curve and may abruptly change direction, perhaps indicating an aerodynamically unstable body. Sometimes thin striae are superimposed on an earlier wide, flat striae system, possibly as a result of turbulent flow.

    B. Ribbed—A form of underdeveloped striation found only on stones. It is intermediate between nodular and striated.

    C. Net—A fine stitchwork structure found on the more friable stony meteorites especially near protuberances.

    D. Porous—A microscopic porosity is often visible on the surfaces of oriented meteorites using a hand lens of between 15 and 30 power. It is found on both irons and stones.

III. REAR SURFACES

    A. Warty—Warty crusts are common on iron meteorites and, in the case of large masses, may form a texture easily visible to the naked-eye. It is very rare on stones and is underdeveloped where it does occur (usually within depressions or at the edges of rear-facing surfaces). It is due to glassy material being released into the dust train and then immediately being drawn back as the meteoroid reaches its retardation point.

    B. Scoriaceous—Characteristic of stony meteorites. Appears as a form of clinkered slag caused by internal frothing at edges and rear surfaces. It is rare on irons, being found only on sharp edges and protuberances.

Table 4.3 summarizes this information.

Table 4.3 Occurrence of Fusion Crusts		
Crustal Class	Irons	Stones
Close Textured	Yes	No
Nodular	Rare	Yes
Striated	Yes	Yes
Ribbed	No	Yes
Net	No	Yes
Porous	Yes	Yes
Warty	Yes	Rare
Scoriaceous	Rare	Yes

4.2.2 Regmaglypts (Fig. 4.3)

Regmaglypts (or regmaglyphs) are millimeter to centimeter grooves, pits, and thumb-like indentations found on the surfaces of meteorites.

Regmaglypts occur on the primary ablation surfaces of irons in regular wavy or reticulated patterns. In some cases large circular pits are to be found, especially on the rear, as a result of the ablation of troilite nodules.

Fig. 4.3: *The Haig meteorite, found in Western Australia in 1951, shows a very well-preserved regmaglypt structure. It is now on display at the Western Australian Museum in Perth. (Courtesy Western Australian Museum)*

Stony meteorites show a wider variety of regmaglypts. Anterior fusion crusts tend to be smooth and quite thin, but on the rear the crust thickens and turns rough, sometimes developing a network of wavy structures. Generally, radiating or sub-parallel grooves are found on the exteriors of stony meteorites. These grooves tend to be concentrated near the corners on the anterior of conical, oriented specimens where the flow lines curve over an abrupt edge of the rear-facing surface. They are normally elongated near the front of the meteorite and rounded towards the back, but are never found on the leading face.

4.2.3   Oriented Meteorites

Many meteorites are rounded or elongated like a football—a sign that they have tumbled during their passage through the atmosphere—but some are oriented with a smooth pointed nose and a flat, sometimes rough rear. Oriented meteorites, therefore, either did not spin during their flight, or the axis of rotation was parallel to the meteorite's trajectory (Fig. 4.4).

Fig 4.4a: *Example of an oriented meteorite, Nuevo Mercurio, Mexico. Note also the contrast in color between the interior and the fusion crust. (Courtesy Smithsonian Institution, Museum of Natural History)*

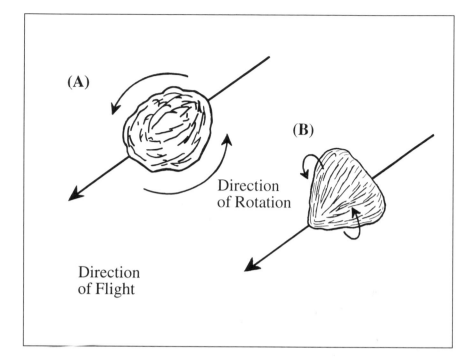

(A)

(B)

Direction
of Rotation

Direction
of Flight

Fig. 4.4b: *A meteorite which has a spin axis at right angles to the direction of flight will become rounded as each side is ablated (A). Where the spin axis lies along the flight trajectory (B), or where the meteorite does not rotate, only one hemisphere is ablated and becomes pointed. In this case the meteorite is said to be* oriented.

4.3   Internal Features

If a meteorite is broken in half, the way in which it has been structured is sometimes evident in the form of chondrules, inclusions, fissures, etc. Grinding the cut face until it is smooth, using progressively finer abrasives, eventually results in a highly polished surface which reveals far greater detail. With irons, further treatment with acids or heat is usually necessary, but all of these methods are within the ability of most amateur collectors (see Chapter 6). Preparing "thin sections" for viewing through a microscope, however, is very much the forte of the professional, but ready-prepared sections are sometimes available through geology groups, and photomicrographs can be obtained from a number of meteorite traders (see Appendix C).

4.3.1   Breccias

Breccias are meteorites composed of broken rock fragments—called clasts—cemented together by finer-grained material. They sometimes display a "shattered" or "crazed" appearance and can be aggregates of widely different rock fragments that have become compacted during impact events. This latter type, known as regolith breccias, represents about half of all carbonaceous chondrites. Amphoterites often display prominent brecciation.

Brecciation can occur in most types of meteorites and often results in weak, friable structures. There are two principal classes of breccia:

1. Monomict, in which the clasts are of the same material as the more finely-grained surrounding host. Many achondrites fall into this category.

2. Polymict, in which the clasts are not closely related—if at all—to the host material. The clasts are often angular in shape. Polymict breccias are rare in irons.

Brecciation may occur at two stages in the history of a meteorite:

1. On the parent body during crustal or mantle movements (monomict) or because of the impact of another body (polymict).

2. While the meteoroid is in space as the result of a collision with a second meteoroid (polymict).

A.V. Jain and M.E. Lipschutz have estimated that some 65% of all iron meteorites have been subjected to shock pressures in excess of 130 kb, and G.J. Taylor and D. Heyman note that only 12 out of 103 ordinary chondrites they examined had not been appreciably shocked. The structural effects of shock and brecciation are listed in Table 3.4; the chemical changes can be found in Table 3.3.

4.3.2   Chondrules (Figs. 4.5 and 4.7)

In 1802—before meteorites were recognized as being extra-terrestrial—a researcher referred to one stony meteorite as containing "curious globules." When Gustav Rose catalogued the Berlin collection in 1864, he named the globules chondrules—from the Greek *chondros* meaning "grain of seed"—and those meteorites containing chondrules came to be known as chondrites.

Chondrules are mainly high-temperature minerals and glasses that are generally spheroidal or ellipsoidal in shape. Identical structures are unknown

Fig. 4.5: *One of the Allende meteorites showing chondrules on the broken face. Also see Fig. 4.7. (Courtesy Smithsonian Institution, Museum of Natural History)*

in terrestrial rocks, the closest equivalent being spheres of augite glomerocrysts found in a basaltic dolerite dyke at Mt. Belches, Western Australia.

Chemically, chondrule bulk composition varies, but except for a somewhat lower metal content, their average composition is essentially the same as that of the host matrix material. Some are predominantly metallic and can contain other opaque minerals. Others contain interstitial glass that may be clear and translucent but which, more commonly, is turbid, indicating devitrification. There are primarily two sorts of chondrules: monosomatic, composed of a single crystal; and polysomatic, composed of several crystals which may be of a single or multiple mineral species.

Fig. 4.6a: *Photomicrograph of a radiating pyroxene chondrule from the Clovis meteorite. The pyroxene can be seen as narrow lines set in a glassy background.*

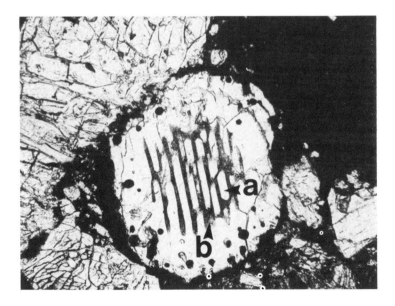

Fig. 4.6b: *Photomicrograph of a barred olivine chondrule from the Ngawi meteorite.* **a** *= olivine (light areas),* **b** *= glass (dark areas).*

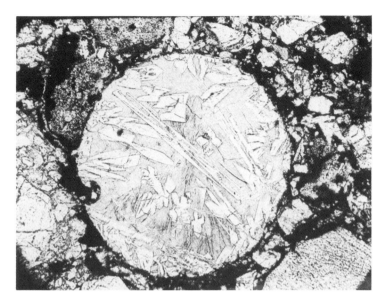

Fig. 4.6c: *Photomicrograph of a glassy chondrule from the Ngawi meteorite.*

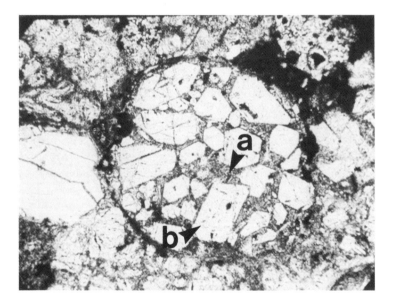

Fig. 4.6d: *Photomicrograph of a porphyritic chondrule from the Clovis meteorite.* **a** *= glass (mottled areas),* **b** *= olivine (light areas). (Photographs courtesy Dr. Brian Mason, Smithsonian Institution, Museum of Natural History.)*

Chondrules show a very limited range of size within each meteorite group, with the CV meteorites containing the largest chondrules and EH the smallest. The sequence is usually

$$CV > LL > L > H \sim CM \sim CO \geq EH$$

Chondrules frequently account for 70% of the mass of a meteorite, sometimes more. They are thought to have originated in space early in the history of the Solar System by the melting of pre-existing solids. They show a variety of textures including overlapping, concentric trachytic alignment of olivine laths, alignment of crystal grains parallel to the border of the chondrule (suggesting solidification in droplet form), and moulding, indicating plastic characteristics when forming. Six main types of chondrules are generally recognized (Figs. 4.6a–d):

1. **Barred Olivine Chondrules.** The outer rims contain droplets of mixed metal sulfide with an inner rim of olivine. Interiors contain skeletal olivine crystals arranged parallel to one another with soda-alumina-silica glass or microcrystalline material. Many are monosomatic and show simultaneous extinction when viewed through crossed nicol prisms.

2. **Radiating Pyroxene Chondrules.** These contain radiating laths, 1–$10\mu m$ in width, of Ca-poor pyroxene with occasional interstitial glassy or microcrystalline material.

3. **Porphyritic Chondrules.** This is the most common type of chondrule, showing greater textural variation than those previously mentioned. They consist of olivine and pyroxene crystals, which may be subhedral (i.e., partially bounded by own crystal faces) or euhedral (i.e., completely bounded by own crystal faces), set in a microcrystalline or glassy material. The relative proportions of pyroxene and olivine are infinitely variable. Generally, the pyroxene is monoclinic, which under polarized light shows twinning of lamellae. Occasionally, the pyroxene grains are coated with augite, and within some grains olivine is arranged poikilitically (i.e., a large pyroxene encloses rounded olivine blebs). Chondritic meteorites of Type 3 have zoned olivine and pyroxene grains in which the iron content usually decreases towards the center.

4. **Glassy Chondrules.** These are rare, being found only in Type 2 and Type 3 carbonaceous chondrites, and as the name suggests, consist almost entirely of glass rich in soda, alumina, and silica. They often contain feathery, skeletal, radiating, or euhedral crystals of pyroxene, olivine, or spinel ($MgAl_2O_4$).

5. **Dark-Zoned Chondrules.** The concentration of opaque minerals (e.g., metal and sulfides) in these chondrules increases towards the surface;

hence the center is lighter than the outer layers, although they are generally dark overall.

6. Lithic Fragment "Chondrules." Probably not true chondrules, these occur in irregular or angular shapes which appear to be fragments of larger crystalline masses. The olivines and pyroxenes are nearly always constant in composition.

Fig. 4.7: *The Allende, Mexico, Type 3 carbonaceous chondrite, which fell February 8, 1969, showing calcium- and aluminum-rich inclusions (white) and numerous smaller chondrules. (Courtesy Smithsonian Institution, Museum of Natural History)*

4.3.3 Calcium- and Aluminum-Rich Inclusions (CAI's) (Fig. 4.7)

CAI's appear as white irregular clasts imbedded within a chondritic matrix. They are especially abundant in certain types of carbonaceous chondrite but also occur in the ordinary chondrites. Most have some sort of concentric structure caused by minerals of differing compositions, and others are similar to chondrules, suggesting that they formed by crystallization from a liquid. A number of CAI's also contain minute nuggets of rare metal alloys and platinum that have been named Fremdlinge, from the German for "little strangers."

CAI's are important because they contain refractory minerals which were the first to condense from the Primitive Solar Nebula and anomalous isotopes which probably originated in a nearby supernova explosion. Hence, they provide an insight into the earliest stages of Solar System formation. They are currently the subject of intensive investigation.

Fig. 4.8: *Etched end-piece of the Bennett County, South Dakota hexahedrite, showing Neumann Lines. The dark streaks to the bottom right are troilite-daubréelite aggregates. Longest dimension about 18cm. (Courtesy Smithsonian Institution, Museum of Natural History)*

4.3.4   Neumann Lines (Fig. 4.8)

Neumann Lines occur in certain iron meteorites and in the kamacite in chondrites. They appear as series of very fine fractures that can be revealed when the specimen is cut, polished, and then etched with acid. They are the product of shock loading at pressures of about 10 kb and at temperatures of less than 575K and are diagnostic of hexahedrites representing cross-sections through very thin plates that penetrate deep into the main mass of the iron. They can be cleaved in three mutually perpendicular directions (i.e., along the faces of a cube or hexahederon) and are named after Johann G. Neumann, who satisfactorily explained their nature in 1848.

4.3.5 Widmanstätten Pattern (Fig. 4.9)

The Widmanstätten Pattern is diagnostic of octahedrite meteorites and becomes visible when the meteorite is cut, polished, and then either etched with acid or gently heated. In some cases the pattern is discernible on weathered surfaces.

The Widmanstätten Pattern consists of broad bands of kamacite sandwiched between narrow ribbons of taenite and arranged parallel to the four pairs of faces of an octahedron (Fig. 4.10). The angle at which the bands intersect on a cut face depends on the orientation of the cut. If the cut is parallel to the face of a cube, then the angle between the lamellae will be 90°; if parallel to an octahedral face, the angle is 60°. During etching the nickel-poor kamacite in the bands is more readily attacked by the acid than the nickel-rich taenite, leading to a relief structure. The width of a set of bands is determined by three factors: nickel concentration (high nickel concentrations inhibits growth), cooling rate, and nucleation temperature.

For a given nickel concentration and cooling rate, material which nucleates at high temperatures will achieve coarser structures than that which nucleates at lower temperatures.

The nickel concentration of taenite is inhomogeneous, decreasing towards the center of the taenite regions. If the nickel concentration is plotted against the distance across the taenite field, a characteristic M-shaped profile develops (Fig. 4.11), indicating that the parent body cooled too rapidly to allow the nickel to diffuse to the center of the fields. Between the kamacite bands are triangular or polygonal areas consisting of a fine intergrowth of taenite and kamacite, which has come to be called plessite, from the Greek for "filling."

The Widmanstätten structure is the result of the parent bodies cooling at between $1° - 100°C/Ma$ at pressures of less than 10 kb and cannot be reproduced in the laboratory. If a meteorite containing the Widmanstätten structure is subjected to prolonged heating at high temperatures, the pattern will be obliterated and cannot be restored.

The pattern is named after Count Alois de Widmanstätten, a Viennese porcelain manufacturer who described the structure in detail in 1808, but it was actually discovered by the Englishman G. Thomson in the Krasno-

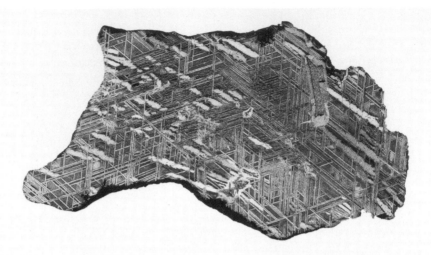

Fig. 4.9a: *The Widmanstätten Pattern of the Edmonton, Kentucky, fine octahedrite. Band width 0.32mm. (Courtesy Smithsonian Institution, Museum of Natural History)*

Fig. 4.9b: *Polished and etched slab of the Jerslev, Denmark iron which is classed as a coarsest octahedrite. Finger-sized kamacite crystals can be seen together with schreibersite and some taenite in the grain boundaries. Note the corrosion at the top of the slice. (Courtesy Dr. Vagn F. Buchwald, Technical University of Denmark)*

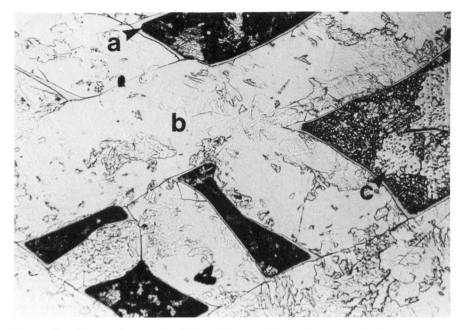

Fig. 4.9c: *Photomicrograph of the Widmanstätten Structure in the Akyumak, Turkey iron.* **a** = *taenite ribbons,* **b** = *kamacite bands,* **c** = *plessite filling.*

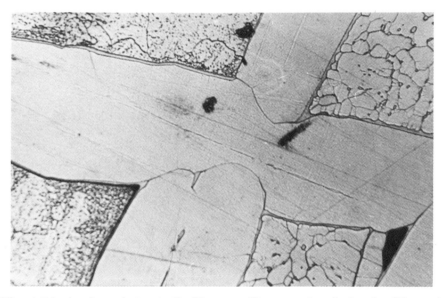

Fig. 4.9d: *As above, but note the Neumann Lines crossing the bands. (Photomicrographs courtesy Dr. Colakoglu, Gazi University)*

Fig. 4.9e: *Detail in the Widmanstätten Pattern. Note how the bands intersect each other. Bar = 1mm*

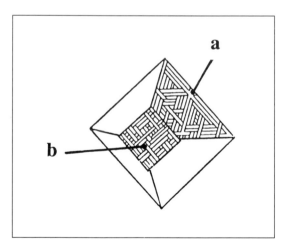

Fig. 4.10: *The angle at which the bars of the Widmanstätten Pattern intersect depends on how the octahedron has been sectioned. If cut along the face of the octahedron (a), the angles will be 60°. Sectioned along the face of a cube, the bands form right angles (b). (Based on Tschermak, 1894)*

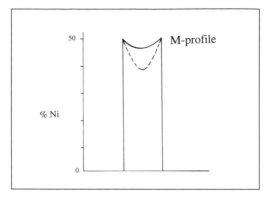

Fig. 4.11 *The nickel concentration in the taenite ribbons of Widmanstätten struc-ture decreases toward the center, producing an M-shaped profile. The solid line represents a moderate cooling rate; the broken line is for a high cooling rate.*

jarsk pallasite. The width of the kamacite bands has been used to categorize octahedrite meteorites (see section 5.5.2).

Fig. 4.12 *Reichenbach Lamellae is extremely rare and very fragile. This example is from the Ilimaes iron which was discovered in Chile during the latter half of the last century. (After Tschermak)*

4.3.6   Reichenbach Lamellae

Reichenbach Lamellae resembles the Widmanstätten Pattern but is caused by the separation of thin troilite plates that have grown on very thin chromite lamellae. The structure is quite rare and fragile.

4.3.7   Stony-Iron Admixtures

Those meteorites that contain substantial amounts of both iron and silicates can produce some of the most striking assemblages. In some specimens hot molten metal has intruded into cooler stony material; in others the reverse appears to have occurred. In a third type, impacts on the parent bodies have resulted in fragments of iron and stone being welded together (see section 5.4).

4.3.8   Albedo

The albedo of a meteorite is the proportion of incident light reflected by the meteorite's surface. It is normally given as a percentage or decimal fraction (e.g., 18% or 0.18) and is principally a function of physical properties, such as surface microstructure and particle size, although the presence of certain minerals (e.g., carbon) can influence the albedo.

The carbonaceous chondrites are the darkest meteorites, reflecting less than 15% of the light incident on their surfaces (usually considerably less), whereas L-chondrites, eucrites, and diogenites have albedos of up to 34%.

Chapter 5

The Classification of Meteorites

5.1 Introduction

The first attempts at classifying meteorites, begun by the 1840's, were based on obvious structural and chemical differences. One of the earliest classifiers was Paul Partsch, Curator of the Vienna collection, who separated the stones from the irons. He then divided the latter into dense, compact irons, and those that contained some stony material in their structure. He also divided the stones into normal and anomalous, sub-dividing the normal specimens further into magnesium-rich and magnesium-poor groups.

In the mid-nineteenth century, Charles U. Shephard developed his own classification system. Like Partsch he had two principal categories of stones and irons but sub-divided the stones into trachytic, trappean, and pumice-like, and the irons into malleable homogeneous, malleable heterogeneous, and brittle. However, Shephard's system was fraught with difficulties which were exacerbated by the fact that, unknown to him, his collection contained several specimens that were not meteorites.

At about the same time that Shephard was attempting to categorize meteorites, A. Boisse was working on the same problems, but he had an advantage over Shephard in that he based his system on petrographical and density factors. His three-class system grouped meteorites into irons, stones, and uncompacted material. The stones were further divided into magnetic and non-magnetic, but the real downfall of Boisse's system was the third class which included viscous or gelatinous matter, colored rainwater, snow, and powders. At that time it was generally believed that meteorites regularly deposited such material.

Carl von Reichenbach dismissed Shephard's system and did not mention that of Boisse in his publications. An abrasive, highly unpopular character, he had been involved in a long-running dispute with Vienna—then the center for meteorite research—and had declared that Partsch's classifications had no sound basis. In 1859 he introduced his own taxonomy based on electro-chemical relationships. His system was, in effect, a measure of the nickel-iron content, although he also expressed the view that a system based on chem-ical composition would have been more desirable—something that was not possible at that time because of the lack of reliable data. Partly because of flaws in his system, but also because of flaws in his character, Reichenbach's classifications found few supporters.

By far the most notable attempt at classifying meteorites in the nineteenth century was that of Gustav Rose at the University of Berlin in 1863–1864. His categories employed mineralogical criteria, and much of the terminology which he introduced is still in use today. Like his predecessors he divided specimens into stones and irons. The irons were sub-divided into almost pure nickel-iron, pallasites (nickel-iron plus olivine), and mesosiderites (nickel-iron plus iron oxide, olivine, and augite). He went on to divide the stones into seven categories: chondrites, howardites, chassignites, chladnites, shalkites, eucrites, and carbonaceous chondrites. His knowledge of meteorites was sec-ond to none since he had been a student of the subject for more than forty years. Highly respected, he also helped to introduce the terms "schreibersite," "troilite," and "Neumann Lines."

In 1872 Gustav Tschermak modified Rose's system, incorporating the newly discovered facts of the nature of bronzite and enstatite in the or-thopyroxene solid-solution series. He further refined the system in 1883. Tschermak's five categories consisted of (1) irons, (2) irons with silicate in-clusions; (3) chondritic-textured stones with olivine, bronzite, and some iron; (4) olivine, bronzite, or pyroxene meteorites; and (5) augite, bronzite, and lime feldspar meteorites with shining crusts. The irons were then sub-divided into octahedrites showing the Widmanstätten Pattern, hexahedrites charac-terized by Neumann Lines, and ataxites that had no discernible structure. He observed that the band widths in the Widmanstätten Pattern varied and thus sub-divided the octahedrites into fine (Of), medium (Om), coarse (Og), and curvilinear (Ok). Tschermak also noticed differences in those irons that had silicate inclusions—the second of his main classes—and thus sub-divided them into pallasites (olivine set in iron), mesosiderites (bronzite and olivine in iron), siderophyres (bronzite in iron), and grahamites (olivine, bronzite and plagioclase in iron). His chondritic group was broadly similar to that of Rose but included nine sub-divisions based on color and texture. In the fourth group, he included Rose's chladnites and chassignites but changed the name of the shalkites to diogenites in honor of Diogenes of Apollonia (see Chapter

1) and added diopside and augite bustites and olivine and bronzite amphoterites. Tschermak's final class included Rose's eucrites and howardites.

In 1863 Maskelyne suggested that meteorites should be termed äeroliths (stones) siderites (irons) and mesosiderites (stony-irons), and four years later C.U. Shephard made another failed attempt, this time to introduce the terms "litholites," "lithosiderites," and "siderites." In France in 1867, Daubrée also worked on a classification system based on iron content. This was later modified by Meunier, but although it was accepted in France, it did not find favour elsewhere.

Between 1885 and 1904, Brezina expanded the Rose-Tschermak categories and introduced the term "achondrite" to describe those stones lacking chondrules. His final system had in excess of 70 sub-groups, many of which contained only one specimen.

In 1920 George Prior revised the Rose-Tschermak-Brezina system. He kept the stones, irons, and stony-irons, and sub-divided the stones into chondrites and achondrites, distinguishing four separate groups based on the ratio of iron to nickel, the type of magnesium silicate, and the form of feldspar that was present. More recently, in 1967, W. Randall Van Schmus and John A. Wood introduced a comprehensive classification system for chondrites which, though subsequently modified, is now universally accepted. At about the same time that Van Schmuss and Wood were working on stones, John T. Wasson at UCLA was re-examining the work of John F. Lovering, who in the late 1950s had found a relationship between the gallium and germanium concentrations in irons. Using improved techniques, Wasson and his co-workers included in their analyses iridium and nickel content, thus developing an entirely new system for classifying irons.

The systems employed today are therefore based mainly on the work of Rose, Tschermak, Brezina, Prior, Van Schmus and Wood, and Wasson, though it would be erroneous to think that the systems are complete or perfect. As new and better investigative methods are developed, and as more and improved data become available, the relationships among the different types of meteorite will become more evident and undoubtedly require further revision and expansion of the classifications. In addition, there are a number of "anomalous" meteorites that do not satisfy the criteria of the present categories, and yet, they must fit into the cosmic jigsaw puzzle somewhere.

5.2 Modified Classification System

The meteorite classifications offered in Table 5.1a are a modified version of the generally accepted categories. Usually, the achondrites are divided into calcium-rich and calcium-poor groups, but such a distinction is of little use to amateur collectors—and not a few professional meteoriticists have noted

Table 5.1a
Modified Meteorite Classification System

STONES			
	Chondrites	Carbonaceous	Types C1-C5
		Enstatite	Types E4-E7
		Ordinary:	
		1) Olivine-bronzite	Types H3-H7
		2) Olivine-hypersthene	Types L3-L7
		3) Amphoterites	Types LL3-LL7
	Achondrites	Angrites	ACANOM
		Aubrites	AUB
		Ureilites	AURE
		HED Sub-group:	
		1) Howardites	AHOW
		2) Eucrites	AEUC
		3) Diogenites	ADIO
		SNC Sub-group:	
		1) Shergottites	AEUC
		2) Nakhlites	ACANOM
		3) Chassignite	ACANOM
STONY-IRONS		Lodranites	LOD
		Mesosiderites	MES
		Pallasites	PAL
		Siderophyre	IVA-ANOM
IRONS		Hexahedrites	
		Octahedrites	
		Coarsest	Ogg
		Coarse	Og
		Medium	Om
		Fine	Of
		Finest	Off
		Ataxites	D

that there is no apparent generic link between meteorites of the same calcium group—whereas the howardites, eucrites, and diogenites, for example, do seem to be related, as are the shergottites, nakhlites, and Chassigny. Similarly, grouping irons on the basis of appearance is more important to collectors than using the Wasson "ABC" system.

5.3 Stones

Meteorites composed largely of silicate minerals are generally referred to as stones or stony meteorites. They represent 94.3% of all observed falls and are divided into two groups, the chondrites and the achondrites. The division was originally made on the basis of whether or not a meteorite contained small spheroids called chondrules, but the distinction is now made on

Table 5.1b
Origin of Classification Names

Class Name	Origin
Achondrite	Without chondrules, see Chondrite
Angrite	After fall at Angra dos Reis, Brazil, January 1869
Ataxite	Greek a-, taxis = without structure
Aubrite	After fall at Aubres, France, 14 September 1836
Chassignite	After fall at Chassigny, France, 3 October 1815
Chondrite	After the presence of chondrules. Greek chondros = grain of seed
Diogenite	After Diogenes of Apollonia
Eucrite	Greek eukritos = easily distinguished
Hexahedrite	After crystal form, hexahedron
Lodranite	After fall at Lodran, now in Pakistan, 1 October 1868
Mesosiderite	Greek mesos = middle; sideros = iron
Nakhlite	After fall at Nakhla, Egypt, 28 June 1911
Octahedrite	After crystal form, octahedron
Pallasite	After the explorer P.S. Pallas
Shergottite	After fall at Shergotty, India, 25th August 1865
Siderophyre	Greek sideros = iron; phyrao = knead
Ureilite	After fall at Novo-Urei, Russia, 4 September 1886

a chemical/mineralogical basis with the confusing result that some classes of achondrites do contain chondrules and some chondrites do not.

Reference is often made to ordinary or common chondrites. In reality there is nothing "ordinary" about these—or any other—class of meteorite, but they are the most abundant type of meteorite, hence the name.

5.3.1 Chondrites—General

Van Schmus and Wood introduced a system for categorizing chondrites in 1967 (Table 5.2). The chondrites were separated into five chemical groups (Fig. 5.1)—enstatite, carbonaceous, high-iron content (olivine-bronzite chondrites), low-iron content (olivine-hypersthene chondrites), and low-iron, low-metal content (amphoterites)—and each group was sectioned into six petrologic types depending on the criteria listed in Table 5.3. The system has now been extended to include seven petrologic types and has been re-interpreted. Whereas in the original system it was generally believed that the degree to which meteorites had been metamorphosed increased from Type 1 through to Type 6, it now seems that Type 3 chondrites are unaltered; those of higher types have been subjected to high temperatures, and those of lower types have been altered by the presence of liquids (aqueous alteration).

The beauty of the Van Schmus-Wood system is that chondritic meteorites can be referred to quite specifically by using a simple shorthand notation. Thus, a C3 meteorite is a carbonaceous chondrite of petrologic Type 3.

Table 5.2 Van Schmus-Wood Chondrite Classification System						
Aqueously Altered		Unaltered	Increasingly Metamorphosed			
1	2	3	4	5	6	7
Carbonaceous (C)						
	Olivine-bronzite (H)					
	Olivine-hypersthene (L)					
	Amphoterites (LL)					
	Enstatite (E)					

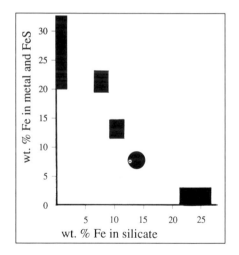

Fig. 5.1a: *Relationship between reduced iron (metal and sulfide) and oxidized iron (silicate) in the five different chemical groups. E = enstatite, H = high iron content, L = low iron content, LL = low iron and low metal content, and C = carbonaceous.*

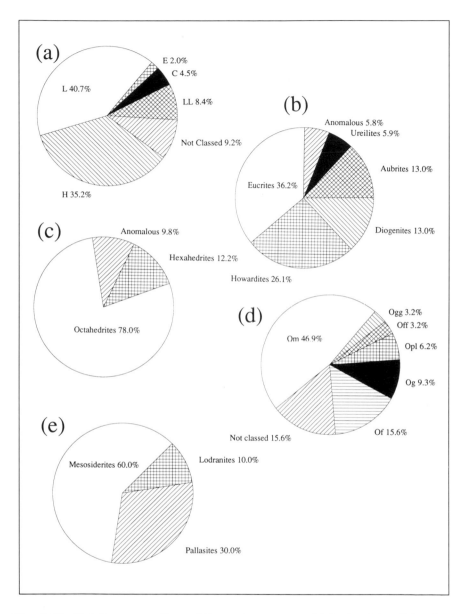

Fig. 5.1b: *Relative proportions of the various groups of* **(a)** *chondrite,* **(b)** *achondrite,* **(c)** *various groups of iron meteorites (Note that, to this date, no ataxites have ever been observed to fall.),* **(d)** *octahedrite, and* **(e)** *various groups of stony irons. (All data based on falls listed in the* Catalog of Meteorites, 4th Edition.)

Table 5.3
Criteria for Distinguishing Different Petrologic Types of Chondrites
(Based upon Van Schmus and Wood, 1967)

	Aqueously Altered	Unaltered		Increasingly Metamorphosed			
			Petrologic Type				
	1	2	3	4	5	6	7
I. Homogeneity of olivine and Pyroxene compositions	—	→ Becoming more homogeneous →					
II. Structural state of low-Ca proxenes	—		Mainly monoclinic			Orthorhombic	
III. Degree of development of secondary feldspar	—	Absent		Mainly microcrystalline aggregates		Clear interstitial grains	
IV. Igneous glass	—		Clear, isotropic primary glass of variable abundance	Turbid if present		Absent	
V. Metallic minerals (maximum Ni content)	—	Taenite <20%		Kamacite and Taenite > 20%			
VI. Ni content of sulfide minerals	—			→ Decreasing Ni content →			
VII. Chondrules	Absent		Sharp, well-defined	Well-defined	Readily delineated	Poorly defined	
VIII. Matrix	Fine, opaque		Opaque	Transparent microcrystalline		Recrystallized	
IX. Bulk carbon content				→ Decreasing from 5% to <0.2% →			
X. Bulk water content				→ Decreasing from 22% to <1.5% →			

Table 5.4 Typical H-Chondrite Composition	
	wt%
SiO_2	34.5
MgO	22.0
FeO	9.7
Al_2O_3	2.8
CaO	1.5
Na_2O	1.0
Cr_2O_3	0.7
P_2O_5	0.5
H_2O	0.5
MnO	0.4
C	0.2
K_2O	0.1
TiO_2	0.1
Fe	18.3
Ni	1.6
Co	0.1
FeS	6.0

Meteorites are regarded as "chondrites" when the abundances of their non-volatile elements are similar to those found in the sun. With the exception of some carbonaceous chondrites, most chondrites contain variable amounts of nickel-iron metal, and all contain an iron-bearing sulfide. (Compare this composition to terrestrial surface rocks that have negligible sulfide contents and no metal.) It is found that those chondrites richest in olivine are poorest in metal, and vice versa. A typical chondrite is 90% silicon, oxygen, magnesium, and iron, and in the most abundant type of chondrite—the olivine-bronzite or H-Chondrites—these elements combine to form the minerals given in Table 5.4.

The chondrules that for so long characterized the chondrite class of meteorite are set in a variety of matrix minerals. The two most commonly found kinds are minute grains of olivine and pyroxene together with minor sulfides, oxides, and feldspars, and graphite mixed with iron oxide. Chondrules often account for more than 70% of the mass of a chondrite and are found in all types except Types 1 and 7.

Chondrites can be as porous as sandstone, having a total pore space of 7 to 18% (sandstone 14%), suggesting that they have not been subjected to the sort of high pressures encountered at great depths. This is one of the main reasons why meteoriticists regard them as surface rocks. Although some chondrites are inherently strong (the La Lande, New Mexico L5 chondrite has a compressive strength of 3810 kg/cm^2), many are quite friable and can be easily crushed between finger and thumb (e.g., Bjurböle).

Apart from chrondrules some chondrites also contain calcium- and aluminum-rich inclusions which appear as white, irregular shaped clasts displaying concentric structures (see Chapter 4.3.3). A number of chondrites are regolith breccias. These are characterized by light-colored xenoliths imbedded within a dark, clastic matrix, and are thought to be representative samples of the parent body regolith that has evolved through aeons of meteorite bombardment. When recovered soon after a fall, most chondrites display a matt-black fusion crust, which is quickly destroyed by weathering mechanisms. Generally, chondrites are dark, the so-called "black chondrites" reflecting only about 4.5% of the light incident on them.

5.3.2   Enstatite Chondrites

Enstatite chondrites are quite rare, representing less than 2% of all stony meteorites. They contain little oxygen, a fact which has led some authorities to suggest that they may have formed closer to the sun than most other types of meteorites. There are, however, problems in transferring a body from a Mercury- or Venus-like orbit into one that intersects the Earth, but proponents of the view have argued that this unusual orbital shift may be one of the reasons why enstatite chondrites are so rare.

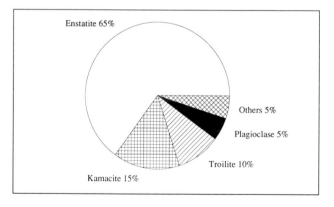

Fig. 5.2: *Composition of E-Chondrites*

As their name suggests, the enstatite chondrites are composed mainly of enstatite (Fig. 5.2), which is very close to pure MgSiO$_3$. Because they are so highly reduced, nearly all the iron occurs as metal or in combination with sulfur to form troilite. Calcium, titanium, and chromium are also found as sulfides. In some cases even potassium and sodium sulfides have been recorded—combinations that are almost unknown on the oxygen-rich Earth. An excess of silica prevents the formation of olivine but allows the growth of tridymite and cristobalite, both of which are polymorphs of silica. Another member is quartz, which is found only in this type of meteorite. The free metal in enstatite chondrites (representing 20–28wt%) is entirely kamacite in composition and is similar to the alloy found in hexahedrite irons. Diopside exsolved from the enstatite occurs as a minor accessory mineral.

5.3.3 Olivine-Bronzite Chondrites

Olivine-bronzite chondrites are the most abundant class of meteorites. They are more oxidized than enstatite chondrites and contain the most perfect chondritic fabrics. Glass is also found in this type of meteorite, though only in Type 3.[1]

H-Chondrites, as they are otherwise known, are composed of approximately equal amounts of olivine (with 15–20 mol.% Fe_2SiO_4) and bronzite together with about 16–21% free metal of octahedrite composition. Clinobronzite, troilite, and the feldspar oligoclase are common accessories, accounting for up to 10% of the composition. The total iron content is normally 27 wt%, with nickel-iron representing 12–20 wt%. Density is 3.6–3.8 gm/cm^3.

5.3.4 Olivine-Hypersthene Chondrites

The L-Chondrites, the second most abundant class of meteorites, are even more oxidized than the H-Chondrites. Olivine-hypersthene chondrites contain olivine with 21–26 mol.% Fe_2SiO_4 and orthopyroxene (commonly called hypersthene) together with about 7–12% free metal. Troilite and oligoclase are found as accessory minerals ($<10\%$). Under a microscope a calcic clinopyroxene is visible. The nickel-iron is a mixture of taenite and kamacite and represents 5-10 wt% of the meteorite. The "L" refers to the low-iron content of 23 wt%.

5.3.5 Amphoterites

The amphoterites were once classed as achondrites. They contain an unusually ferroan olivine (27–32 mol.% Fe_2SiO_4), which shows intense recrystallization. They also have less than 7% free metal, which is nickel- and cobalt-rich.

The "LL" (Tables 5.1 and 5.2) refers to low-iron (20 wt%), low-metal (2 wt%).

---

[1]Glass can also be formed by shock.

5.3.6   Carbonaceous Chondrites (Fig. 4.7)

Despite their name some carbonaceous chondrites actually contain very little carbon. They form a more heterogeneous group than do the other classes of chondrites.

Carbonaceous chondrites are rich in oxygen, so they contain little or no free metal, since most of it has combined with the oxygen and silicon. They are generally composed of mixtures of minerals that were formed at different temperatures. Types 1 and 2 have undergone aqueous alteration from fluids circulating through the rock at low temperatures and transforming matrix material into an array of hydrous phyllosilicates. These are extremely complex, layered structures that are very similar to terrestrial clays.

The carbonaceous classes are also important in that they contain organic matter in the form of complicated and sometimes very large molecules composed of carbon, hydrogen, oxygen, and nitrogen. Such organic material occurs as rings (aromatic hydrocarbons), straight or slightly branching chains (alkanes), and carboxylic and amino acids. These molecules do not appear to be of biogenic origin.

Type 1 carbonaceous chondrites consist of amorphous hydrous silicates, sulfates, and low-temperature mineral assemblages. They also contain a highly magnetic, finely divided nickel-iron spinel but little iron sulfide and little or no free metal. Olivine occurs only rarely. Type 1 carbonaceous chondrites are of low density ($\rho = 2.2$ gm/cm^3) and have bulk carbon contents of 3–5%.

Type 2 carbonaceous chondrites contain sheet minerals, serpentines, chlorite, and other water-bearing, low-temperature assemblages. Some free sulfur is usually present. The paucity of metals results in only weakly magnetic, and in some cases, non-magnetic meteorites with low densities ($\rho = 2.5$–2.9 gm/cm^3). Chondrules are sparse but very sharply defined and are composed of almost pure forsterite, enstatite, or clinoenstatite.

In Type 3's, abundant, distinct chondrules—some of which contain clear, fresh glass—are set in a fine, opaque, olivine-rich matrix in which the pyroxene crystals, when present, are poorly arranged. The chondrules of Type 4 are also abundant and distinct and, like Type 3, the matrix is olivine-rich and fine-grained. However, the matrix is transparent and microcrystalline. The Type 5 carbonaceous chondrites contain chondrules that are indistinct and set in a coarse granular matrix.

Four sub-groups have been identified—CI, CM, CV, and CO—and named after those meteorites that typify the groups. Details of their occurrence are given in Table 5.5.

Fig. 5.3a: *Frontal surface of the Pasamonte, New Mexico, eucrite.*

Fig. 5.3b: *Rear surface of Pasamonte, from the Nininger Collection. (Courtesy Smithsonian Institution, Museum of Natural History)*

Table 5.5
Occurrence of
CI, CM, CV and CO Sub-groups in
Carbonaceous Chondrites

Name	Abbrv.	Occurrence
Ivuna	CI	Type 1 only
Mighei	CM	Type 2 only
Vigarano	CV	Types 2-5 (CV2–5)
Ornans	CO	Types 3-4 (CO3–4)

Only a few CI meteorites are known. They are recognized by the absence of chondrules and an abundance of low-temperature, largely hydrous minerals containing up to 20% water. Their only high-temperature mineral is olivine, which occurs as tiny, isolated, well-formed crystals. Epsomite is found in veins through CI meteorites, possibly (since it is easily dissolved in water) deposited by liquid water flowing through the chondrite parent body. Serpentine and magnetite are also present.

The CM chondrites are the most abundant of the four sub-groups, and like the CI meteorites, they are composed of a matrix of hydrous serpentine minerals together with small amounts of magnetite, epsomite, and other low-temperature minerals. The most obvious difference to the CI chondrites is that they contain chondrules, isolated olivine grains, and rare crystalline aggregates of high-temperature minerals. CM2 chondrites contain only about half the amount of water found in CI meteorites. It has been argued that CI and CM chondrites may be fragments of cometary nuclei rather than asteroidal debris. Comet nuclei are low-density structures that appear to be basically chondritic in nature. When close to the sun, the nuclei begin to disintegrate, producing a stream of dust which, if it encounters the Earth, causes a meteor shower to appear in the atmosphere. Some authorities believe that the high velocities attained by cometary debris are sufficient to destroy most of the chondritic meteorites imbedded within the stream and would therefore account for the low number of observed falls of the CI and CM sub-groups. However, the evidence is by no means conclusive. So far none of the meteorites whose orbits have been accurately determined appears to be associated with any of the known comets, and velocity measurements suggest that most meteorites, regardless of type, enter the atmosphere at less than 30 km/s.

CV chondrites have a fine opaque olivine matrix containing calcium-, titanium-, and aluminum-rich minerals. CV2 chondrites have water-bearing assemblages, and CV3 chondrites often exhibit large chondrules that contain sulfur and metal. The ratio of chondrules to matrix is low. The Allende CV3 has 5–10% CAIs but also includes low-temperature feldspathoids (e.g., sodalite $3NaAlSiO_4 \cdot NaCl$).

Like the CV's, CO chondrites contain calcium- and aluminum-rich phases including spinel, melilite, and a solid-solution mixture of gehlenite ($Ca_2Al_2SiO_7$), akermanite ($Ca_2MgSi_2O_7$), and perovskite ($CaTiO_3$). CO3 chondrites contain an abundance of densely-packed tiny chondrules (< 0.2 mm diameter).

5.3.7 Achondrites—General

Achondrites are relatively rare, representing only 7.6% of all observed falls. Although they were originally defined on the basis of lacking chondrules, they have since been re-defined according to their mineralogy and chemistry so that today some categories of achondrites (e.g., howardites) do display rare chondrules. The achondrites are a heterogeneous group that are generally igneous, but some are also mixtures of rock fragments. More coarsely crystallized than chondrites, some achondrites are chemically, mineralogically, and texturally very similar to terrestrial dolerites and basalts. Nearly all enstatite achondrites are brecciated due to the presence of coarse crystals that weaken their structure. Most are monomict breccias. They appear to be the product of melting on small planetesimals, having crystallized 4,500 Ma ago.

Achondrites are often subdivided according to their calcium content—being either calcium-poor or calcium-rich—though there seems to be no direct relationship between meteorites of the same calcium group.

The calcium-poor subdivision includes meteorites with 0–3% Ca. Virtually the entire silicate fraction of such meteorites is composed of MgO and SiO_2. The nickel-iron alloy present is kamacite with the composition of hexahedrites (i.e., 6% Ni). This subdivision includes the aubrites, diogenites, ureilites, and the unique Chassigny. The calcium-rich subdivision incorporates the eucrites, howardites, shergottites, nakhlites, and the single Angrite, and covers calcium contents from 5–25%.

5.3.8 Angrite

Only one Angrite exists, that which fell at Angra dos Reis, Brazil, in January, 1869. Its relationship to other meteorites is unclear but it appears to have formed in a fairly large body.

In effect Angrite is a pyroxene cumulate composed of 90% augite—one of the few meteorites to contain substantial amounts of this mineral—together with some olivine and spinel. Only 1.5 kg of the Angrite fall was ever recovered, and it is not available to amateur collectors.

5.3.9   Aubrites

Aubrites are enstatite achondrites named after the fall at Aubres, France, on September 14, 1836. They are relatively rare and are similar to enstatite chondrites—though the relationship, if any, is obscure. Aubrites are a light sandy color because of their lack of iron (1–2%), and when newly fallen, they have a white or transparent fusion crust. Although some are *very* tough, others are among the most friable of all meteorites and need to be kept in glass jars to prevent fragments from being lost. Small quantities are normally preserved in gelatin cells.

Aubrites are largely composed of enstatite which is very nearly pure $MgSiO_3$, plus a small amount of accessory minerals. Most are brecciated, several being regolith breccias. It would appear that aubrites formed under highly reducing conditions, possibly in the Primitive Solar Nebula or as products of igneous processes on their parent bodies.

5.3.10   Ureilites

Ureilites are coarse-grained olivine-pyroxene (pigeonite) achondrites with carbonaceous materials in the grain boundaries. Mineralogically, they may have formed from Type 3 carbonaceous chondrites, but the relationship is not well understood. Ureilites contain minor amounts of carbon as finely divided graphite and diamond and can have greater amounts of nickel-iron than any other type of achondrite.

Their coarseness suggests a plutonic origin, but it is not yet clear whether they are cumulates of grains from a magma chamber or the residual grains left over from partial melting. The olivine and pyroxene grains show preferred orientations, a fact which suggests cumulation—though some scientists dispute this interpretation. In any event the mixture is difficult to achieve by known mechanisms.

5.3.11   Eucrites (Fig. 5.3)

The eucrites are the most abundant of the achondrites and appear to be related to both the diogenites and howardites. They are meteoritic basalts, some of which are vesicular, and many are monomict breccias. Compared to terrestrial basalts, they are rather more reduced and have a lower content of volatile elements, but they are visually similar. This similarity may be one of the reasons why they are rarely found. They tend also to resemble lunar basalts.

Eucrites are commonly fine-grained with the elongated plagioclase grains being encased in pyroxene. Some have been metamorphosed forming equant, interlocking grains indicative of recrystallization in a solid state, possibly from being buried or re-heated by subsequent lava flows.

Mineralogically, eucrites contain pigeonite with or without hypersthene, calcic plagioclase, olivine or tridymite, ilmenite, chromite, magnetite, troilite, and nickel-iron. Small amounts of free metal are usually found.

Damage to crystals by the solar wind indicates that eucrites formed within the inner Solar System, and some researchers have suggested that the asteroid 4 Vesta may be the parent planet as, spectroscopically, its surface is similar to eucrites.

5.3.12 Diogenites

Related to the eucrites are the diogenites or hypersthene achondrites. They are very similar to igneous pyroxenites, containing a fairly iron-rich bronzite plus small amounts of accessory minerals.

Diogenites display interlocking, plutonic crystals that are larger than those found in the eucrites, and this coarseness means that they are susceptible to brecciation. The pyroxene is more magnesium-rich than that of the eucrites, suggesting earlier crystallization and accumulation.

5.3.13 Howardites

Howardites are the second-most abundant achondrite group. They are basaltic, composed of plagioclase (An_{80-97}) and pyroxene, dominantly hypersthene. Howardites are usually brecciated and are thought to be "soils" of eucrites and diogenites cemented together by impact mechanisms. They also contain a minor chondritic component.

5.3.14 Shergottites

The shergottites, together with the nakhlites and Chassigny, form a separate suite within the achondrite class known as the SNC meteorites. Only eight examples are known (Table 5.6).

Shergottites are basaltic achondrites—in effect calcium-rich eucrites—but their main difference from "normal" eucrites is that they appear to have formed only 1,300 Ma ago. They are composed mainly of a plagioclase, maskelynite, which is more sodic (An_{50}) than that found in either the eucrites or the howardites, and two kinds of pyroxene, calcium-rich augite and calcium-poor pigeonite. They are also unusual in that they, and the other SNC meteorites, are the only achondrites to contain bonded water, shown in the presence of minerals such as kaersutite. The elongated pyroxene crystals are oriented, suggesting they accumulated at the bottom of a magma chamber.

Table 5.6
SNC Suite of Meteorites

Type		Location and Date	Notes
Shergottites	Shergotty	Shergotty, India, 25 August 1865	Single stone fell, 5kg.
	Zagami	Zagami, Nigeria, 3 October 1962	Single stone fell, 18kg.
	EETA 79001	Elephant Moraine, Antartica, Dec 1979–Jan 1980	Single stone found, 794.2gm.
	ALHA 77005	Allan Hills, Antartica, 1977	Single stone found, 482gm.
Nakhlites	Nakhla	Nakhla, Egypt, 28 June 1911	40 stones fell, total mass of 40kg.
	Lafayette	Lafayette, Indiana, USA, known prior to 1931	Single stone, 800gm, found in mineral collection of Purdue Univ.
	Governador Valadares	Governador Valadares, Brazil, 1958	Single stone found, 158gm.
Chassignite	Chassigny	Chassigny, France, 3 October 1815	One or more stones fell, 4kg.

The origin of the shergottites is puzzling because of their age—they are far too young to have originated on the asteroids or on the Moon, lunar volcanic activity having ceased over 3,000 Ma ago. This fact has led some scientists to speculate on a Martian origin for the SNC suite. Their arguments are based on the fact that Mars was an active planet until quite recently, and studies have shown that an oblique impact by a small asteroid would be sufficient to launch fragments of the Martian surface into orbits that would eventually intersect that of the Earth. In addition SNC meteorites contain certain noble gases in the same concentrations as those measured in the Martian atmosphere, which is strong supporting evidence for this theory.

5.3.15   Nakhlites

The second type of meteorite in the SNC suite is the nakhlites. These are pyroxene-olivine cumulates which, texturally, are locally ophitic and similar to terrestrial dolerites. Nakhlites are composed of 75% diopside and 15% olivine. Iron- and titanium-oxides are also found together with some water.

Like the shergottites, they are very young and may have a similar origin. They are unshocked.

5.3.16 Chassignite

Only one chassignite exists, that which fell at Chassigny, France, on October 3, 1815. It is a calcium-poor olivine achondrite very similar to terrestrial dunite. The olivine accounts for 90 wt% and forms chondrules. A nickel-rich NiFe alloy is also present. Similar to the shergottites and nakhlites, Chassigny is young (1,300 Ma) and only mildly shocked.

5.4 Stony-Irons

Stony-iron meteorites are a fairly minor class, representing little more than 1% of all observed falls, but they are, nonetheless, an important group that includes some of the most attractive meteorites known. They are often seen as a transitional class between the stones and irons, but this is a gross oversimplification, as the histories of the various types of stony-iron meteorites appear to be quite different.

For the purposes of this book, the stony-irons have been separated into four main groups, but one—the single siderophyre—is not normally recognized in current literature.

5.4.1 Lodranites

Only two lodranites exist; the first fell at Lodran, which is now in Pakistan, on October 1, 1868; the second was recovered from the Yamato foothills, Antarctica, in 1979 and is known as Yamato 791493. Consequently, they are rarely, if ever, offered for sale.

Lodranites are coarse, friable aggregates of approximately equal amounts of olivine (Fa_{13}) and pyroxene (Fs_{17}) grains which are associated with a discontinuous subordinate aggregate of nickel-iron (9.4% Ni). Grain centers are iron-rich.

5.4.2 Mesosiderites (Fig. 5.4)

Ingraciously referred to as "wastebuckets," the mesosiderites are the second most abundant of the stony-irons. They are fragmented polymict breccias which display a greater diversity of matrix than other brecciated meteorites.

Mesosiderites contain almost equal amounts of silicates and metals. The silicate fraction is broadly analogous to the HED achondrites and is composed of fine grains forming an igneous or metamorphic structure. Typically

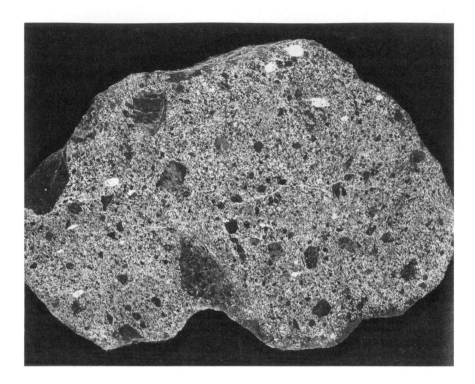

Fig. 5.4: *The Crab Orchard (Tennessee) mesosiderite, discovered in 1887. This cut and polished slab reveals the brecciated nature of the meteorite. It was once classed as a "grahamite." Neg. Nr 314903. Photo: Thane Hierwert. (Courtesy Department Library Services, American Museum of Natural History)*

they include hypersthene or bronzite, plagioclase, clinopyroxene, pigeonite, olivine in large clasts, and tridymite (the high-temperature polymorph of SiO_2). The nickel-iron metal does not form a continuous network but consists of fine grains and occasionally centimeter-size nodules. It has a composition similar to that of the octahedrites (7.1–10% Ni), and suitably large nodules show Widmanstätten Patterns when polished and etched.

Mesosiderites were probably formed by shock mechanisms deep within their parent bodies, resulting in the mixing of metal with silicate. They may be associated with the eucrites. In 1978 R.J. Floran developed a classification system for the mesosiderites according to the nature of the silicate matrix (Table 5.7).

Table 5.7 Modified Floran Classification of Mesosiderites	
Type	
1	Little recrystallization of matrix.
1A	Plagioclase abundant.
1B	Plagioclase deficient.
2	Moderately recrystallized matrix.
2A	Pyroxene-poikiloblastic matrix.
2B	Plagioclase-poikiloblastic matrix.
3	Highly poikiloblastic matrix.
3A	Pyroxene-poikiloblastic matrix.
3B	Plagioclase-poikiloblastic matrix.
4	Intergranular (melt rock) matrix.
4A	Plagioclase abundant.
4B	Plagioclase deficient (not represented).

5.4.3 Pallasites (Fig. 5.5)

The pallasites are the most common—and for many the most attractive—of the stony-iron meteorites. They consist of olivine crystals set in a continuous network of nickel-iron formed by hot, liquid metal being forced upwards into a cooling, contracting, and cracking mass of olivine.

The olivine grains are often single crystals up to 1 cm in diameter and are normally magnesium-rich. The iron is of octahedrite composition and may show the Widmanstätten Pattern when polished and etched. In some examples large areas of silicate-free iron are found. Troilite, schreibersite, and farringtonite are present as accessory minerals. Brian Mason notes that the pallasites can be divided into two groups, as detailed in Table 5.8.

5.4.4 Siderophyre

Only one example of a siderophyre exists, that which was found at Steinbach, Germany, in 1724. Officially it is classed as an "IVA—Anomalous" stony-iron. The term "siderophyre" is now rarely used. Steinbach is a coarse textured stony-iron, similar to many pallasites except that the silicate fraction is composed of an orthopyroxene (Fs_{20}).

5.5 Irons

Iron meteorites are now classed according to the Wasson system, which is based on the concentration of gallium, germanium, iridium, and nickel. How-

Fig. 5.5: *Pallasite from Brenham, Kansas. This slab has been cut, polished, and etched to reveal the Widmanstätten Structure of the nickel-iron component. Note how the olivine crystals are set in the metal and do not form a continuous mass. Neg. Nr. 314889. Photo: Charles H. Coles (Courtesy Department of Library Services, American Museum of Natural History)*

Table 5.8				
Mason's Pallasite Groups				
Group	NiFe	Ni	Olivine	Fa Range
1	55%	10%	Fa_{13}	Fa_{10-16}
2	30–35%	15%	Fa_{19}	Fa_{16-21}

ever, from a collector's point of view, it is easier to use the traditional method of classification according to structure. Indeed, most catalogs of meteorites continue to use this system as well as that of Wasson. Iron meteorites are considerably denser than rocks found on the Earth's surface, a fact which helps with their identification.

5.5.1   Hexahedrites

Hexahedrites are easily recognized by the presence of Neumann Lines which become apparent when a specimen is cut, polished, and etched with acid. They are invariably heavy, having a density of about 7.9 gm/cm^3.

Hexahedrites mainly consist of large cubic (hexahedron) crystals of kamacite. Their iron content is usually about 93.5%, nickel 5.5%, and cobalt 0.5%. Traces of phosphorus, sulfur, chromium, and carbon are often found. Rhabdite crystals are sometimes associated with the iron sulfide troilite.

A gradation from hexahedrite structure to coarsest octahedrite has been observed in some irons involving localized segregations of taenite on the margins of kamacite crystals. This feature may indicate a common origin with the octahedrites.

5.5.2 Octahedrites

Octahedrites, like the hexahedrites, are also easy to identify, this time because of an intricate network of kamacite bands known as the Widmanstätten Structure (see Chapter 4.3.5). They contain schreibersite, cohenite, chromite, and diamond together with nodules of graphite and troilite, which are often found walled-in by kamacite. A classification system for the octahedrites has been developed based on band width. (Table 5.9)

Table 5.9 Octahedrite Sub-groups				
Type	Abbrv.	Band Width (mm)	$\approx$Ni%	Notes
Coarsest	Ogg	>3.3	6.4–7.2	Rare. Taenite sometimes present.
Coarse	Og	1.3–3.3		Common.
Medium	Om	0.5–1.3	7.4–10.3	Most common.
Fine	Of	0.2–0.5		Common.
Finest	Off	<0.2	7.8–12.7	Rare. Distinct bands of kamacite.
Plessitic	Opl	<0.2		Kamacite sparks and spindles.
Ataxite	D	No pattern	>10.8	Well-developed, slowly annealed plessite. Kamacite spindles very rare.

5.5.3 Ataxites

Ataxites were once divided into two groups depending on their nickel-content, but this system has since been abandoned. They obtain their name from the Greek *a-, taxis* meaning "without order," as originally researchers were unable to detect any internal structure. However, apart from those that have a nickel content in excess of 25% (and which are therefore composed entirely of taenite), a microscopic structure is visible which shows discrete taenite crystals fringed with laths of kamacite but without intersection of lamellae. The kamacite laths are set in a fine plessite matrix which forms a very significant proportion of the entire mass. Most ataxites contain less than 20% nickel and appear to have a common origin with the other classes of irons.

Chapter 6

The Finer Points of Meteorite Collecting

6.1 Introduction

Once a meteorite has been purchased, it is necessary to preserve it and make it presentable for exhibiting, and when several meteorites have been acquired, it is advisable to catalogue the collection accurately in order to avoid possible identification errors at some future date. If people are aware of your interest, it is quite likely that you will, from time to time, be presented with all manner of rocks which they believe to be meteorites. So how are meteorites identified, preserved, catalogued, and exhibited?

$$\boxed{\text{WARNING}}$$

Some of the procedures described in this chapter involve the use of hazardous chemicals and abrasives. If you have limited laboratory experience, you should seek the help of a professional, and everyone should follow these few simple rules:

1. Protect Yourself! Always wear a lab coat, goggles, and rubber gloves when using chemicals and always handle chemicals with care. Keep your work area well-ventilated.

2. Protect Your Family! Always lock chemicals and abrasives away out of the reach of children and never put chemicals in soda pop bottles. Ensure that neither children nor pets can enter the room in which you are working.

3. Protect Your Work Area! Cover all work surfaces with protective sheeting. Wipe up any spillages immediately and dispose of unwanted chemicals carefully.

IF IN DOUBT SEEK PROFESSIONAL HELP!

6.2   The Identification of Meteorites

Meteorite identification can sometimes be very easy, but it can also be very nearly impossible without the advantage of a well-equipped laboratory. Familiarity with the tell-tale signs is important, however, especially in the field, and often a few pertinent questions and one or two careful observations are all that are necessary to tentatively identify a suspect specimen. The following routine should help:

1. First, ask why the inquirer thinks the object is a meteorite. Did she or he see it fall? Did it form a crater? Was it warm when it was found? Were there reports of a possible meteorite fall in the area? Did the object cause any damage? Simply finding a rock surrounded by broken glass in one's conservatory does not necessarily mean that the object is a meteorite.

2. Feel the weight of the object. If it is very heavy, it may be an iron meteorite.

3. Look for the following external features:

   (a) A fusion crust

   (b) Oriented shape

   (c) Lack of sharp edges (Fig. 6.1)

   (d) Flow lines

   (e) Regmaglypts

   (f) Fractures

   (g) Widmanstätten Pattern on weathered surfaces.

4. Meteorites that have been exposed to the Earth's environment for some time will have been eroded by weathering mechanisms. Look for signs of rust and, if a crater exists, check for iron shale.

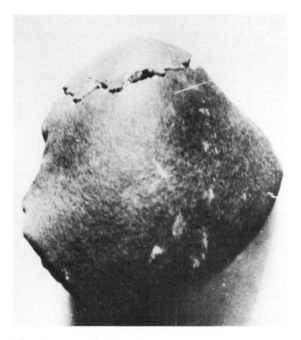

Fig. 6.1: *The Gujargaon, India, H5 chondrite (fell 4 September 1982). Note the lack of sharp edges and the damage to the meteorite. (About 25cm diameter.) (Courtesy Director General, Geological Survey of India)*

5. If possible examine the interior for chondrules, nickel-iron, and contrasting color to the exterior.

6. Perform the following simple tests:

(a) If a fusion crust appears to be present, scrape it to see how deep it is. It should be no more than 1–5 mm.

(b) Use an emery wheel to grind away a small part of the object and examine for small bright flecks of metal.

(c) Crush a small amount in a mortar and test for magnetic particles.

(d) Test for nickel, using the following procedure:

    i. Dissolve a small amount of the metal (about 2 mm diameter) in about 3 cc of concentrated hydrochloric acid.

    ii. Add a few drops of nitric acid and boil for about 2 minutes.

    iii. Add a drop of citric acid and allow the solution to cool.

    iv. Neutralize by carefully pouring in 15 ml of ammonia.

v. Add a few drops of dimethylglyoxime solution in alcohol. If nickel is present a bright scarlet precipitate will appear.

If the object does seem to be a meteorite, it should be brought to the attention of the national or local natural history museum, university, or geological survey office. They will probably want to see the specimen, but if the owner is unwilling to part with it, only about 20 gm will be needed for a positive identification.

6.3   Preservation

Meteorites are never recovered in pristine condition. Exposure to the damp, oxygen-rich atmosphere with its pollution and to the equally hazardous Earth results in meteorites quickly becoming contaminated and weathered. Obviously, the sooner a meteorite can be recovered after falling the better, but most specimens lie for some considerable time before being salvaged and suffer as a consequence. Needless to say, those meteorites that land in arid areas—such as deserts and in the Antarctic—are the least affected, and researchers in Antarctica now handle those meteorites recovered on that continent in much the same way that precious lunar rock is processed.

Cleaning meteorites is the first step in their preservation. Newly-fallen meteorites should always be given to a professional to clean in order to minimize the risk of damage, but "old" specimens can be cleaned by anyone who has the patience, as it is unlikely that an amateur could do more damage to a specimen than the environment has already inflicted. The exceptions are those specimens that are extremely friable and which even the experts find difficult to clean.

6.3.1   Dust

All meteorite samples not contained within air-tight cases require regular dusting at least once per week. If left too long, layers of dust will accumulate which may be difficult to remove, especially from the more porous meteorites (e.g., chondrites).

Dusting is best done with a soft camel-hair brush. Photographic shops stock brushes that have a bellows mechanism attached which can be used to blow the dust off—certainly a better method than using one's breath, which invariably introduces moisture into the specimen.

Start by brushing the underside of the specimen and then slowly rotate it so that it is always the underside that is being cleaned. This procedure ensures that the dust falls off the specimen and onto the floor and is not simply re-distributed over the meteorite's surface.

Thick layers of dust can be removed by scrubbing with a hard brush and anhydrous alcohol. A toothbrush is often sufficient, but nylon ones should be avoided as they can leave marks on the specimen. Some damage to meteorites during scrubbing is unavoidable, so be prepared to lose a few chondrules. Providing the specimen is not too friable, ultrasonic generators, which should release most of the dust, can be applied for 1–5 minutes. Of course, prevention is better than cure, and keeping specimens in air-tight boxes and display cabinets avoids many of the problems caused by dust.

6.3.2 Clay

Many finds are often covered in layers of clay that can usually be removed with a hard brush without too much trouble. If the layers are too thick, immersing the sample in water or vinegar can often do the trick. They should then be washed in distilled water and bathed in alcohol for two or three days.

6.3.3 Rust

Rust (iron oxide) is one of the main problems facing meteorite collectors. The first signs are a red, brown, or yellow discoloration of the sample. If caught at this stage, the damage to the specimen will be only superficial, but left untreated, an iron meteorite can rapidly deteriorate into little more than a pile of loose fragments. Rust can be removed with sodium citrate, sodium hydrosulfate, or oxalic or tartaric acid. Small individual stains are best treated using a cotton ball soaked in a solution of one part sodium citrate in five or six parts water. The ball should be applied to the stain for 10–20 minutes.

After treatment for rust the sample should be immersed in an alcohol bath for a couple of hours and sealed using a good lacquer (see section 6.3.7). In some cases it may be best to cut out the rusted area.

6.3.4 Deliquescence

The meteoritic mineral lawrencite readily absorbs moisture from the air, leading to rapid erosion of those irons containing it. The mineral is said to be deliquescent and is a nuisance to collectors, who often refer to it as "meteoritic cancer." It is also present in pallasites and some stones.

Lawrencite is ferrous chloride ($FeCl_2$). When exposed to moist air, it changes to ferric chloride, ($FeCl_3$), and limonite, an iron hydroxide. A proportion of the ferric chloride is then reduced to its lower oxidation state, but the rest undergoes further changes, reacting with water vapor to produce more limonite, and hydrochloric acid, which dissolves the meteorite. Its presence is revealed by the occurrence of tiny, green, viscous droplets that are usually found between the plates of the Widmanstätten Pattern.

Unfortunately, there isn't much that can be done about lawrencite, other than keeping the afflicted specimen in a dry nitrogen-rich atmosphere—and even this may not work! An alternative that sometimes works is to put the meteorite into a plastic zipper bag or air-tight jar together with a sachet of water-absorbing silica gel crystals. These can be obtained from most drugstores, but usually only in 1 kg bags! As only a few grammes are required, it is necessary to transfer some of the crystals into a smaller porous packet. Some collectors use cloth pouches, but an equally good method is to use a tea bag—with the tea leaves first removed, of course!

6.3.5   Light

Meteorites do not generally contain minerals that are particularly photosensitive, but the colors of all minerals will gradually fade when subjected to intense light, especially strong sunlight. Unfortunately, a high level of illumination is required to show certain features to their best advantage, but providing specimens are kept in the shade while not on display, they should maintain their appearance for a considerable time. Avoid daily exposure to bright sunlight to minimize cracking.

6.3.6   Cracking, Scratching, and Fragmentation

Meteorites that are friable are likely to crack and scratch quite easily and must be handled with care. Extremely fragile specimens are best kept in colorless glass jars so that they can be viewed without having to be touched. Fragments should be preserved in gelatin.

6.3.7   Lacquering

Lacquer, if correctly applied, protects specimens from dust, rust, oxidation, and humidity. First, ensure that the sample is dust-free (see section 6.3.1) and that any rust has been removed (see section 6.3.3). Immerse the meteorite in an anhydrous alcohol bath for two or three hours in order to remove any trapped moisture, and allow to dry. Repeat two or three times with fresh alcohol. Using one of the commercially available lacquers, spray the meteorite with a thin layer, ensuring that the lacquer does not run. Allow to dry according to the manufacturer's recommendations before applying a second coat. It is important to keep the meteorite as dust-free as possible while drying.

Through time the lacquer may become discolored, or rust may develop where moisture has been trapped in the meteorite, in which case the lacquer must be removed and, if necessary, the meteorite treated before being re-sprayed. Fortunately, most good lacquers can easily be removed with acetone. The usual procedure is to use a lint-free cloth soaked in the solution and rubbed gently against the meteorite's surface, working the solution into any pores or fissures. The sample is then prepared and lacquered as above.

6.4 Display Methods

When dealers do not supply meteorites ready for display, they must be prepared. This is not a particularly difficult task, though it can be time consuming. However, a well-prepared specimen, properly lit, makes the effort all worthwhile.

6.4.1 Cutting and Polishing

In most cases the internal structure of a meteorite is best revealed by cutting and polishing a slab. This is especially true when it is desirable to show the number, shapes, sizes, and distribution of chondrules, and is essential when one wishes to display the Widmanstätten Pattern or Neumann Lines. However, not all meteorites need to be cut, and where a specimen has a good oriented shape, such treatment is nothing short of vandalism! In addition, the most friable meteorites are virtually impossible to cut and polish successfully.

Cutting a meteorite can take some time, especially in the case of irons in which the nickel-iron mixture creates a particularly tough alloy. Consequently, sectioning stones is best left to someone equipped with a diamond-tipped saw. These are often found in natural history museums, universities, and in industries dealing in the manufacture of metals and the utilization of rocks and minerals. In recent years there has been an increase in small jewellery-making businesses, some of which have invested in diamond-tipped saws. A number of meteorite traders are willing to cut specimens, and full details of their services are given in their brochures. The best idea is to ask around. Some establishments will gladly help, and some do not even charge.

Iron meteorites are best cut using a hacksaw rather than a diamond-tipped saw, as the nickel-iron alloy tends to loosen the diamonds from their mountings at great expense!

When sectioning a meteorite, it is often best to cut as large a face as possible in order to show the maximum amount of detail. This is particularly the case with the coarsest octahedrites, where the Widmanstätten Structure is so large as to be unrecognizable on small surfaces. Iron meteorites, especially, should be kept cool by running distilled water into the cut in order to avoid the possibility of thermal alteration of the internal structure. The meteorite should then be dried with acetone and gently warmed on a hot-plate.

Once the specimen has been cut, it is necessary to get rid of the gouges left behind by the saw and to polish the surface to a high luster. A set of four or five abrasive carborundum sheets should be purchased in various grades from coarse to very fine and attached firmly to a flat work top so that they cannot slip. Using the coarsest grade first, slowly but firmly rub the cut face of the meteorite back and forth across the abrasive sheet. Continue until the marks left behind by the saw have been removed. Using the next finest grade, repeat the procedure until the scratches from the coarse sheet have been smoothed away. Repeat, using progressively finer abrasive sheets and always rubbing at right angles. Finally, polish the surface with a soft lint-free cloth or buffer attached to a power tool. It is usually at this stage that one notices small fragments of carborundum imbedded in the specimen. These can usually be removed with a fine-tipped bradawl or needle. In order to avoid slipping with the bradawl and scratching the polished surface, clamp the specimen in a vice and use both hands to guide the tool. A bright light and desk-top magnifying glass, which can be bought in combination, are useful in this procedure.

Some meteorites are corroded near their exterior surfaces, and many collectors remove the damage in order to make the specimen look more attractive and to minimize the possibility of the corrosion spreading. Small areas can be drilled out using an appropriate size bit; larger sections will have to be removed using a hacksaw. Remember at all times to keep the specimen cool.

6.4.2   Etching

Once a stony meteorite has been cut, polished, and lacquered it is ready to be put on display. Irons and stony-irons require etching, however, which involves handling acids and other hazardous chemicals. **Extreme care should be exercised at all times during this procedure. Etching should be done with the assistance of a professional and under a ventilation hood. The etching solution is called "nitol" in the U.S., and "nital" in the U.K.**
The following materials are required:

1. 30 cc concentrated nitric acid

2. 95% (190° Proof) ethyl alcohol (ethanol)

3. Distilled water

4. Glass dish

5. 2 glass beakers

6. Paint brush

Proceed with the following steps:

1. Pour 30 cc of concentrated nitric acid into one of the beakers.

2. Pour 500 cc of ethyl alcohol into the other beaker.

3. Pour the acid slowly into the alcohol while stirring the alcohol beaker. Handling these chemicals should only be done by individuals protected by laboratory aprons, rubber gloves, and eye protection. Remember, if in doubt seek professional help. NEVER STORE THIS MIXTURE—IT CAN EXPLODE!!!

4. Once the solution has been mixed, pour it into the glass dish.

5. Hold the meteorite over the dish and brush the solution onto the polished surface. Always brush away from yourself into the dish in order to avoid being splashed.

6. After less than one minute, the Widmanstätten Pattern should appear. Continue etching until a satisfactory structure appears. (Note that it is possible to over-etch and destroy the pattern, in which case it is necessary to re-polish the specimen.)

7. When etching is complete, rinse the meteorite in distilled water thoroughly.

8. Place the meteorite immediately in an alcohol bath for a couple of hours to remove any remaining moisture.

9. Allow to dry and then coat with lacquer (see section 6.3.7).

Some collectors display the etched and untreated halves side-by-side to show the difference. Others etch only a portion of the polished surface.

Etching meteorites can have its problems, the most common of which follow:

1. During the alcohol bath brownish stains appear along fissures and near inclusions. These stains can be removed with silver polish and a soft lint-free cloth. Rinse with distilled water and place in a fresh alcohol bath for a couple of hours. This procedure may have to be repeated several times.

2. During etching the structure turns dull and flat, showing little or low relief. When this occurs the meteorite has been over-etched. Wash in distilled water and then dry. Re-polish the cut face and begin etching again. Successful etching comes only with experience. Determine the optimum etching time by experimenting with one half of the meteorite and use this as a guide when treating the other half.

3. After etching, no pattern emerges. This can stem from several causes:

   (a) The acid might be too weak. Add a few drops of nitric acid reagent solution.

   (b) The meteorite might not be an octahedrite or hexahedrite, it could be an ataxite.

   (c) The pattern may have been destroyed because the meteorite has been artificially heated or heavily shocked.

   (d) The specimen might not be a meteorite at all. Check for nickel (see section 6.2).

4. The lacquered surface becomes discolored or dull. If this happens within a few days of etching, then the lacquer used is unsuitable and an alternative must be found. Nearly all lacquers will discolor with age, in which case they will have to be removed with lacquer thinner, acetone, or alcohol. Dampen a soft cloth with the lacquer remover and rub gently but firmly across the cut surface.

5. Rust develops on the cut surface. This is caused by moisture which was trapped by the lacquer or has entered through a scratch or deterioration of the lacquer layer. Remove the lacquer as above, clean out the rust, and re-polish. If this is a recurring problem, perhaps because of the presence of lawrencite, keep the meteorite in an air-tight container with a small sachet of silica-gel crystals to absorb the moisture (see section 6.3.4).

6.4.3   Heating

Another way of bringing-out the Widmanstätten Pattern is to gently heat the meteorite over a naked flame until the "temper" colors are revealed. Do not, however, subject the meteorite to high temperatures for long periods, as this will destroy the structure.

6.4.4   Exhibiting Specimens

Whether one intends to exhibit a meteorite privately or publicly, it will command much more attention if attractively displayed. A number of mineral traders now stock a variety of display stands and cases (Fig. 6.1).

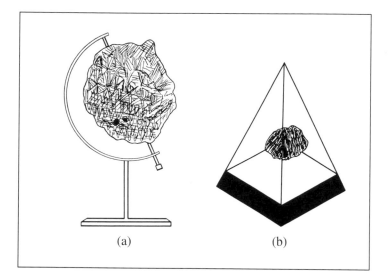

(a)   (b)

Fig. 6.1: *Exhibition stands.* **(a)** *Caliper Stand.* **(b)** *Lucite Pyramid.*

Adjustable caliper and double-caliper stands are quite popular and are available in 24 kt. gold-plated brass. Easel stands hold specimens in much the same way as an easel holds a blackboard. Peg stands consist of three upright pegs, on which the meteorite rests. The problem with all these methods is that the specimens collect dust and therefore require regular cleaning (see section 6.3.1). In order to avoid this problem, some collectors keep their meteorites and stands in display cabinets or place the meteorites in display cases. A number of meteoritophiles encase their meteorites in clear crystalline Lucite. Kits, which consist of a resin, a hardener, a coloring agent, and a set of mold can be obtained in most model shops. They are inexpensive, easy to use and attractive but have the disadvantage that once the Lucite has been allowed to set, the meteorite cannot be removed.

Lighting, another important aspect of exhibiting meteorites, can make or mar a display. Although excessive and prolonged illumination can fade some minerals (see section 6.3.5), meteorites are not particularly photosensitive. Spot lights are best but tend to get rather hot, so if they are to be used in a confined space—as in a display cabinet—they must be properly ventilated. Attention should be paid to the way in which the light is angled in order to maximize its effect and reduce glare. This is particularly true for irons, where correct lighting can emphasize lamellae, different types of metals, minerals, and inclusions.

Pallasites, if sectioned thin enough for the olivine crystals to become transparent, are best lit from behind, passing the light first through a white perspex (Plexiglass) diffusing screen (Fig. 6.2).

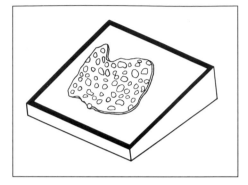

Fig. 6.2: *Pallasites are best lit from behind. Light boxes used by photographers for transparencies and negatives are ideal for this purpose.*

Particularly friable meteorites should be kept in glass jars and fragments preserved in gelatin cells. Very small pieces of meteorites should be exhibited in transparent boxes which incorporate magnifying lenses in their lids (Fig. 6.3).

Although most meteoritophiles are happy simply to display meteorites in their own homes, some enjoy exhibiting specimens to a much wider public. Consequently, they put their meteorites on display at meetings of their local astronomical or geological society or loan specimens to libraries, museums, town halls, and even to those banks whose managers have grown weary of advertising interest rates all the time!

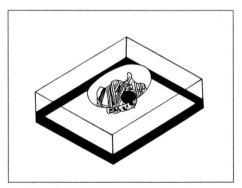

Fig. 6.3: *A magnifying box has a lens built into the lid and is ideal for examining small specimens.*

6.4.5 Labelling

The labelling of specimens is one area to which amateur collectors need to pay more attention. Most specimens are poorly labelled, and many carry no means of identification.

There are basically two ways in which a meteorite can be labelled. The first is to attach a self-adhesive sticker onto the most uninteresting face of the specimen, which carries an identification number linking it to a catalogue of specimens which each collector should keep. The alternative is to paint the number directly onto the meteorite. There are pros and cons in each method. Adhesive labels can usually be removed with little effort, which is useful if the specimen is to be re-catalogued or sold. Unfortunately, labels also have a habit of losing their adhesiveness and falling off. Through time some of the adhesive may diffuse through the paper, making it impossible to read the reference number, and inks often fade, causing the same problem. Painted reference numbers do not normally fade, but neither can they be removed if necessary without scratching the meteorite. There is also the possibility that some paints may react with the meteorite's minerals, but the likelihood is remote.

When exhibiting specimens, more detailed labelling is required, usually in the form of a plaque attached to the display stand. The type of information that should be given depends on the audience. For a group of astronomers or geologists the label might read as follows:

> Odessa, Texas, iron meteorite.
> Coarse octahedrite. Found 1922.
> Note Widmanstätten Structure.

For the public a more general comment is required. For example:

> Iron meteorite from Odessa, Texas.
> This meteorite is at least
> 700 million years older
> than the oldest rocks on Earth.

Labels should be used to identify and inform, but do not fall into the trap of trying to include too much information.

6.5   Taking a Closer Look

Meteorites are composed of aggregates of minerals. The various types of minerals and their arrangement in the structure of the meteorite is best seen at high magnifications.

6.5.1   Meteorites Under a Hand Lens

A relatively inexpensive and simple tool that is often overlooked by novice collectors is the hand lens. These are available in various magnifications, with 10–15x being the most popular. It is worth paying a little bit extra for a good quality lens, first because the lens surfaces will be coated to protect them against scratching, and, secondly, because the lens will have been more correctly "figured," to prevent distortion of the image. Most good lenses come with a case or pouch, though this may have to be purchased separately. Try to ensure that the lens is kept clean at all times by using a soft cloth or camel hair brush, obtainable from photographic stores and opticians. Avoid touching the lens with your fingers and don't breath moisture onto it.

The normally hidden internal structure of a meteorite is best revealed when it has been cut and polished (see section 6.4.1). Of the stony meteorites the chondrites are often the most interesting. Some carbonaceous chondrites—such as Allende—contain chondrules which range from vivid orange to jet black and which are bordered by pyroxenes and feldspars set in a fine-grained carbonaceous matrix. Other chondrites contain flecks of metal between the silicate minerals, and chondrules that become detached leave behind tell-tale circular cavities.

The Widmanstätten Structure in irons is particularly worthy of investigation with a hand lens. If correctly etched the thin taenite ribbons which border the thicker kamacite bands should just be visible. Many irons have inclusions of silicates or of other metals. Odessa often contains small yellow or green olivine crystals, and circular, bronze-colored troilite inclusions surrounded by brittle, silvery schreibersite are found in a number of irons. Black graphite nodules with troilite and schreibersite borders are not uncommon.

Fusion crusts take on a new dimension when viewed through a hand lens. Note how the striae flow across the surface and how they are deflected by lumps of metal or by chondrules. Determine whether the fusion crust is porous and look for the thermally altered zone between the crust and the interior.

In a practical sense, hand lenses are useful for the early detection of rust. Brown, red, or yellow discoloration of the fissures between lamellae is a clear indication that an iron meteorite is beginning to rust and should be treated.

6.5.2 Microscopy

Good microscopes are expensive, but they are sometimes available for use through schools, colleges, universities, or local geology and natural history groups. Binocular and polarizing microscopes are particularly useful.

Used at low power, the microscope is the next logical step up from the humble hand lens. Find out which features are the most interesting through a hand lens; then take a closer look with a microscope. Examine chondrules and CAI's for structural details such as layering and the Widmanstätten Pattern for hair-line Neumann Lines caused by impact-induced shock.

The polarizing microscope is one of the most powerful tools available to the geologist and meteoriticist, but in order to examine specimens with this type of instrument, they must first be carefully and professionally prepared. This involves taking a thin slice of the meteorite and adhering it firmly to a glass slide. When dry, the specimen is ground until it reaches a standard thickness of 0.03 mm, whereupon most of its minerals become translucent.

The polarizing microscope differs from the normal instrument in that it incorporates two sheets of Polaroid, one on either side of the thin section. The lower sheet, between the light source and the thin section, is called the polarizer; the upper sheet is the analyzer, and it can be removed. When light passes through the polarizer only, it is said to be plane-polarized. When the analyzer is inserted, the light is then cross-polarized.

Minerals viewed in plane-polarized light reveal certain properties. Those that appear dark in hand specimens are usually seen in shades of brown or green, while the light colored minerals lose their color altogether (except where the light source is of a low wattage, in which case they may appear pale yellowish, but in reality it is the light source that is colored). If the slide is rotated, the colors of certain minerals may change, a phenomenon known as pleochroism. Some minerals, such as the olivines and pyroxenes, seem to stand out from the mounting medium (resin, etc.) and are thus described as having high-relief. Others (e.g., feldspars), which blend in with the mounting medium, have low-relief. Sets of parallel lines in a mineral indicate cleavage planes.

In crossed-polarized light some minerals suddenly become vividly colored. These are not their true colors, however, but interference colors caused by rays of light being split and rotated by minerals and thus arriving at the analyzer slightly out of phase. Consequently, a mineral may have a variety of interference colors, although some are so strongly colored as to mask

Table 6.1
Properties of the Main Mineral Groups Under a Polarizing Microscope
(*) Denotes Most Important Diagnositic Properties

Mineral	Plane-Polarized Light				Crossed-Polarized Light	
	Color	Pleochroism	Relief	Cleavage	Interference Colors	Twinning
Olivines	None	None, usually	High*	None, usually	Bright yellow, reds, greens and blues	None, usually
Pyroxenes	Colorless to very pale brown	Usually none	Moderately high*	Vertical: one cleavage only Basal: two cleavages at 90°*	Usually bright yellows*, blues, reds and pinks	Rarely observed as simple twins in basal sections
Feldspars	None	None	Low, sometimes appear cloudy	One set of cleavage traces but these are usually faint and difficult to see	Sometimes yellows but more often greys* and white	(i) Simple* (ii) Lamellar* (iii) Cross-hatched (rarely occurs)

the interference. When rotated many minerals will turn black at intervals of 90°—their extinction positions. Minerals that remain black all the time between crossed polars are termed isotropic. The meteoritic glass maskelynite (though strictly speaking not a mineral) is isotropic.

Olivines, pyroxenes, and the rare amphiboles have bright interference colors—vivid reds, blues, and yellows—but feldspars appear only in white, gray, and black. Feldspar is often twinned either as alternative black and white stripes or as cross-hatching caused by adjacent twins reaching their extinction positions at different points.

Table 6.1 lists some of the properties of minerals viewed through a polarizing microscope.

6.6   Keeping Records

Keeping careful records of a meteorite collection is important for several reasons. First, if the collection becomes reasonably large—as collections have a habit of doing—it will not be possible to commit all the necessary details to memory. Secondly, should some of the specimens be offered for sale or exchange, the recipient will probably want some background information; and, lastly, it is advisable to keep records for insurance purposes. Hard copy records are fine, though many collectors now prefer to use computerized files which are quicker to search through, if the program has been correctly structured.

6.6.1   Basic Details

It is up to the individual to decide what information should be kept on file, but consider including the following:

1. **Reference Number.** This should be the same as the Entry Number in the purchase/sales record (see section 2.3) and should be marked on the meteorite. Consecutive numbering is best, though a prefix denoting the type of meteorite is useful (e.g., PAL005 for a pallasite, AUB031 for an aubrite, etc.).

2. **Name of Specimen.** This should be taken from the *Catalogue of Meteorites* (4th Edition) or the *Meteoritical Bulletin* (a supplement to *Meteoritics*). Do not trust the spellings given by vendors!

3. Location.

4. Latitude and longitude.

5. Date on which the meteorite fell or was found, including the time. Always use Universal Time (U.T.), which is the same as Greenwich Mean Time (G.M.T.).

6. Class or type of meteorite (e.g., E5 enstatite chondrite, Om medium octahedrite, etc.).

7. Mass in grams.

8. Description. Friable, polished slab, etched, fusion crust present and type, size, etc.

9. Supplier.

10. Value.

11. Bibliography.

6.6.2 Photography

Some collectors keep photographic records of their specimens, and anyone with a basic knowledge of SLR photography should find little difficulty in doing the same. However, there are a few simple rules to follow.

Specimens should be placed against a plain white background and lit by two lamps (either tungsten or electronic flash) angled at 45° left and right of the incident face. A centimeter ruler or a 1cm block should be included in the photograph for scale. Because most specimens will be relatively small, it may be necessary to use a close-up lens or lens attachment, such as a bellows or extension tube.

Photography through a microscope (photomicrography) is relatively simple, requiring only an inexpensive accessory which attaches the camera body to the microscope eyepiece. Specimens can be lit either by passing light through them—as in the case of a thin section—or by reflecting light from their surfaces. In most cases, the through-the-lens (TTL) metering system found on most modern cameras is sufficient for good exposures. However, where a specimen contains minerals of vastly different colors and contrasts, it is advisable to bracket the exposure by 1–2 stops.

Chapter 7

Tektites

7.1 Introduction

Tektites are dark, glassy objects found only in certain parts of the world. Most are quite small, but specimens up to 15 kg have been discovered. Although they have been known to Man since at least the Aurignacian Period, they have only recently attracted intense scientific interest. They are easier to collect than meteorites because they are more robust and not particularly prone to corrosion. In addition, they are certainly less expensive. Tektites display an interesting variety of colors and shapes which add to their attractiveness.

7.2 History

The first human interest in tektites was shown during the Aurignacian Period by Cro-Magnon Man, who probably used the sharp glassy rocks as simple tools. Splinters dating back to *c.* 29,000 B.C. have been found at Willendorf, Austria, and flaked tools of Libyan Desert glass appear to have been in use between *c.* 10,000–20,000 B.C. Recent studies have also uncovered what appear to be tektite implements from Libya which could be as old as 200,000 years, though they are still under investigation. In the Philippines tools fashioned by pre-Neolithic Man date from *c.* 4,000–6,000 B.C., and after the Iron Age (500 B.C.) tektites were carried as good luck charms. In Indochina they are associated with Neolithic pottery.

The first written reference to tektites was by Liu Sun in China, *c.* A.D. 950, who named them Lei-gong-mo, meaning "Inkstone of the Thundergod." Perhaps not surprisingly, it was Charles Darwin who undertook the first serious scientific study, in 1844, and he came to the conclusion that they were nothing more than terrestrial obsidian (a type of volcanic glass). He regarded

113

the peculiar flanges found on certain tektites as being the result of rapid in-flight rotation.

The first suggestion that tektites were of extra-terrestrial origin was by Victor Streich who, in 1893, expressed the view that they were a form of meteorite. But not everyone agreed with Streich's interpretation, and Verbeek, believing them to be volcanic ejecta, proposed a lunar origin in 1897. Others preferred a more earthly explanation, regarding them as volcanic bombs. Although this theory was popular for a time, it was eventually realized that there were no volcanoes in the vicinity of the tektite areas, and the theory fell out of favour.

At the turn of the century, F.E. Suess became convinced that tektites originated beyond the Earth, and he attributed their peculiar shapes to "sculpturing" by high velocity air flow. In order to test his theory, he built a wind tunnel in which he placed models of tektites and subjected them to a hot airstream. It was at this stage that people began talking of "glass meteorites," though it was Suess himself who introduced the term "tektite" from the Greek *tektos* meaning "molten."

The meteoriticist F. Berwerth rejected the concept of glassy meteorites. He discovered that tektites showed a close chemical similarity to certain sedimentary rocks, in particular sandstone, and suggested they were the products of an ancient human culture.

During the past century the controversy over the origin and nature of tektites continued unabated. The fact that they do not contain cosmic ray tracks indicates that if they have been subjected to the cosmic environment the period was very short—less than 900 years—indicating an origin within the Earth-Moon system. Some researchers, like Nininger, believed that tektites were the ejecta from lunar impacts, whereas others, such as John O'Keefe, considered lunar volcanoes to be the source. On the other hand, many scientists, such as the late Nobel laureate Harold C. Urey and tektite expert Virgil Barnes, took the stance that tektites were essentially terrestrial, caused by the impact of one or more asteroids with the Earth. The return of lunar samples by Apollo XI in 1969 showed that tektites are not associated with the Moon and effectively killed the lunar hypothesis.

In recent years researchers have turned to deep sea sediments, hoping to determine the full extent of the tektite fields and the date of their deposit from the myriads of microtektites that litter the ocean floors.

7.3   Locations

Unlike meteorites, tektites are not scattered randomly across the globe but instead are located in fairly well-defined areas known as strewn fields, seven of which are generally recognized (Fig. 7.1). By far the largest is

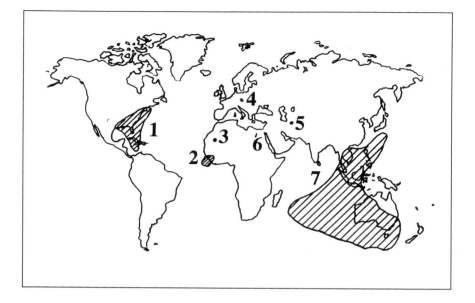

Fig. 7.1: *Map showing the location of the seven principal textite strewn fields: 1. North American, 2. Ivory Coast, 3. Aouelloul, 4. Czechoslovakian, 5. Irgiz, 6. Libyan Desert, and 7. Australasian.*

the Australasian Strewn Field. It stretches from about mid-way between Australia and New Zealand Northwards through the Philippines and South-East Asia and then Westwards across the Indian Ocean almost to the island of Madagascar off the coast of Africa. The tektites are about 750,000 years old.

The second largest area is the North American Strewn Field. Triangular in shape, it covers most of the Caribbean, reaches Northwards to Martha's Vineyard and West to Texas. At about 35 million years old, it is the most ancient of all the strewn fields.

Next comes the Ivory Coast Strewn Field, a roughly East-West oval starting in the Ivory Coast of West Africa and ending somewhere in the mid-Atlantic. It is about 1 million years old. Other locations include Aouelloul Crater, Africa, the Libyan Desert, Irgiz in the Soviet Union, and the Moldavite Strewn Field in Czechoslovakia. Further details can be found in Table 7.1.

Table 7.1
Strewn Fields and Tektite Names

Strewn Field	Tektite Names	Location	Age (Ma)
Australasian	Australites	Australia	0.7
	Billitonites	Billiton Is. (Bilitung Is.)	
	Darwin Glass[1]	Tasmania	
	Indochinites	Indochina	
	Indomalaysianites	Malaya	
	Javanites	Java	
	Muong-nong	S.E. Asia	
	Philippinites	Philippines	
	Rizalites	Philippines	
	Thailandites	Thailand	
Czechoslovakian	Bottlestones	Bohemia	15
	Moldavites	Moldau River	
	Netolice	Bohemia	
	Vltavines	Moravia	
N. American	Bediasites	Texas	35
	Georgiaites	Georgia	
Aouelloul		Mauritania	3(?)
Ivory Coast		West Africa	0.9–1.0
Libyan Desert		North Africa	28
Irgiz	Irgizites	Zhamanshin, USSR	?

[1]Not strictly a tektite

7.4   Appearance

Although some tektites are quite large—about the size of a football and weighing 15 kg—a vast majority are fairly small, being only a few centimeters in diameter with a mass of less than 100 gm. They are usually black but can also occur in shades of brown, gray, and green. Some Czechoslovakian specimens are referred to as "bottlestones" because of their bottle-green coloration. A list of the colors of the various types of tektite can be found in Table 7.2.

Table 7.2
Colors of Tektites

Type	Color
Aouelloul	Greenish/yellowish-gray
Bediasites	Light olive-brown
Darwin glass	Gray
Georgiaites	Shades of olive green
Moldavites	Shades of olive green
Netolice	"Poisonous" bottle green
Vltavines	Brown
Zhamanshinites	Bluish

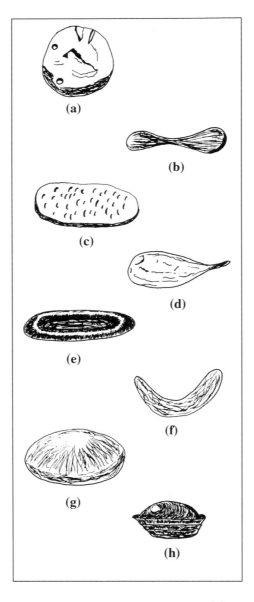

Fig. 7.2: *Various tektite shapes. From top to bottom:* (**a**) *sphere,* (**b**) *dumbbell,* (**c**) *disc,* (**d**) *teardrop,* (**e**) *hollow,* (**f**) *bent, or boomerang,* (**g**) *oval,* (**h**) *button.*

Tektites occur in a variety of shapes, though some strewn fields are characterized by one particular sort. The most common type of tektite is the so-called splash-form, which includes spheres, teardrops, dumbbells, etc. (Fig. 7.2). They often display corrosion markings of three sorts:

1. Cupules; hemispheric pits of less than 1 mm diameter;

2. Gouges; elongated depressions, U-shaped in cross-section, usually with sharp edges; and

3. Meandrine Grooves; again, U-shaped in cross-section but with the look of worm tracks in the wood or sometimes star shaped (Fig. 7.3).

Fig. 7.3: *Star-shaped "worm tracks" are found on the surfaces of some tektites.*

These forms of corrosion—once believed to be a form of sculpturing by primitive Man—are, in fact, the result of hypersonic airflow over the tektite's surfaces and possibly are also due to reactions with surrounding rock.

Flanged buttons are another form of "sculptured" tektite quite popular with collectors (Fig. 7.4). Found mainly in Australia, the core is lens-shaped and surrounded by a flange covered with ringwaves (i.e., concentric rings and double spirals) on the anterior surfaces. The posterior is quite often corroded.

In South-East Asia block-shaped tektites called Muong Nong are found. They are unusual in that they have a layered structure unknown in other forms of tektite.

Finally, many strewn fields contain flattened fragments that have spalled off larger pieces because of thermal shock. They can be quite sharp and need to be handled with caution.

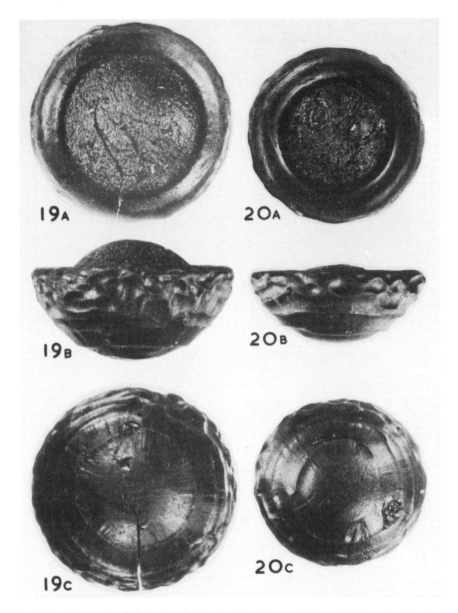

Fig. 7.4a: *Two sets of photographs by the late George Baker of his "perfect button" Australite tektites. Note the flanges and glassy appearance. (Courtesy Dr. John A. O'Keefe)*

Fig. 7.5: *Sculptured thailandite, about 25mm diameter.*

Fig. 7.6: *Pitting on the surface of a tektite from Thailand. About 30mm diameter.*

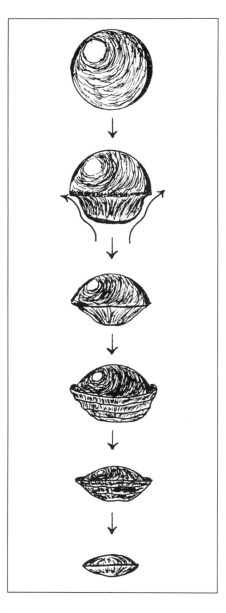

Fig. 7.4b: *Flanged buttons are caused by hypersonic airflow over tektite surfaces. In extreme cases the tektite is reduced to a thin lens. (Based on Baker, 1956)*

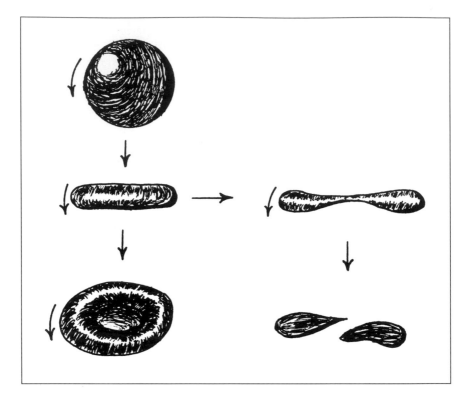

Fig. 7.4c: *Some tektites may have been shaped by rapid rotation in a low-gravity environment.*

7.5   Structure and Composition

While some tektites (i.e., Muong-Nong types) are layered, the splash forms reveal no structural details even when seen in thin section. There are no inclusions, they contain only a few vacuoles (bubbles), and they show a total lack of crystals, a fact which suggests rapid cooling. For lndochinites the cooling rate has been estimated to be less than 10 hours. In teardrop tektites the vacuoles mimic the shape of the section of tektite in which they are found. Hence, those located in the head of the tektite tend to be spheroidal while those in the tail are elongated ellipsoids. Obviously, the tektites have become stretched and frozen while in flight, thus preserving the shapes of the vacuoles.

The way in which a tektite has been subjected to strain can be observed by placing a whole transparent specimen in a dish of light machine oil set between crossed polars. In spherical tektites a black cross appears within the specimen. The cross is caused by the crust cooling first, followed by the interior which then contracted thus pulling the interior into a state of tension.

In composition tektites are somewhere between granite and basalt, similar to andesite but deficient in volatiles (Table 7.3). They are in many ways similar to certain sedimentary rocks. They resemble terrestrial obsidian and can easily be mistaken for it, as Darwin discovered. However, when subjected to a blowpipe, the water in obsidian causes it to froth whereas tektites simply turn molten because they are extremely dry—they contain only 100 ppm of water compared to most igneous rocks which contain 5,000 ppm.

			Table 7.3			
	Comparison Between Tektites and Principal Terrestrial Rocks[1]					
	Tektites	Granite	Basalt	Andesite	Peridotite	Sandstone
SiO_2	73.09	71.0	49.1	57.0	45.2	82.0
Al_2O_3	12.60	13.6	16.9	15.4	2.7	8.3
$FeO + Fe_2O_3$	5.12	4.4	10.4	7.8	7.6	4.4
MgO	2.16	0.8	7.6	6.7	41.4	–
CaO	2.31	2.1	11.6	7.8	2.6	–
Na_2O	1.45	3.5	2.7	2.8	0.3	0.5
K_2O	2.40	4.2	0.2	1.7	0.1	4.8
TiO_2	0.87	0.4	1.5	0.8	0.1	–

Tektites have densities of between 2.21 gm/cm^3 (Libyan Desert glass) and 2.8 gm/cm^3 (bottlestones) and a hardness of 6–7 on the Mohs Scale. By comparison, quartz has a density of 2.6 gm/cm^3 and a hardness of 7.

7.6 Collecting Tektites

Most of the information given in previous chapters on collecting meteorites applies equally to tektites. They have several advantages over meteorites, however, in that they are less expensive, more easily cleaned, and do not require any special treatment to make them look attractive.

Those mineral traders dealing in meteorites invariably stock tektites. Like meteorites, they are usually priced in dollars per gram but tend to be less expensive (Table 7.4). Most specimens are relatively clean when obtained from traders and require only a quick polish with a soft duster to make them

---

[1]**Granite:** Igneous originating from the consolidation of magma found in continental masses. **Basalt:** Igneous, similar to granite, but of greater density and making-up most of the sea bed **Andesite:** Igneous, mid-way between granite and basalt. Found in mountain ranges such as the Andes from which its name is derived. **Peridotite:** Material from which the Earth's mantle is composed. **Sandstone:** sedimentary rock of diagenetically converted sands.

ready for display. When found in their natural state, however, they can be covered in soil, mud, sand, vegetation, etc., and may require cleaning.

Table 7.4 Tektite Prices (1989)	
Type	Price Range (US$/gm)
Australites	1.70– 2.50
Bediasites	1.50– 5.50
Billitonites	8.50–10.50
Indochinites	0.50– 2.00
Javanites	8.00–10.00
Libyan Desert	2.50– 8.00
Moldavites	3.00– 5.00
Rizalites	3.00– 5.00
Zhamanshinites	1.00– 3.00

Crusts of clay can usually be removed with a hard brush. If this fails, try ordinary vinegar or a bleaching solution. One method is to boil the tektite in a concentrated solution of sodium sulfide and then, when cool, wash in lukewarm water. Organic substances, such as lichens, should be treated with aqueous ammonia. Stubborn encrustations can normally be removed with sulfuric acid, but care is required.

Pits in the surfaces of tektites often contain the most resistant deposits and may need to be cleaned using a pin or fine bradawl. It is a long, laborious job which requires a free-standing magnifying glass and a powerful light source. Because tektites are relatively hard, they are not easily damaged, but they should still be handled with some care as there is always some risk.

Tektites are best displayed in a fashion similar to pallasites, with the light source behind the specimen and the light passed through a diffusing screen.

7.7   The Origin of Tektites—An Aside

Tektites are among the most enigmatic objects known to Man. Their limited distribution, composition (especially their dryness), and their lack of cosmic ray tracks have all been difficult to explain, but during the past seventy years several theories have been advanced. These have included a collision between a meteorite and a natural Earth satellite known as a "cosmolith" (V. Goldschmidt, 1921, 1924), dehydration of siliceous gels by humic acid (E.W. Eason, 1921), the ablation of light meteoritic metals (H. Michel, 1922, 1939), a ring of material around the Earth (T.W.E. David, 1927, and K. De Boer, 1929), meteorite impacts (L. J. Spencer, 1930's), and a meteorite impact on the Moon (H.H. Nininger, 1940's). Even antimatter has been blamed by researchers such as M.A.R. Khan.

Planetary scientist John O'Keefe has studied tektites in considerable detail over a number of years. His conclusion was that they were ejected by lunar volcanoes into geocentric orbits forming a series of rings around the Earth. The rings, he argued, lowered the temperature of the Earth by about 20°C, leading to what paleontologists refer to as the Terminal Eocene Event, a mass extinction some 35 million years ago. Unlike terrestrial steam volcanoes, O'Keefe envisaged hydrogen-driven volcanoes sufficiently powerful to launch material to speeds in excess of the lunar escape velocity.

Although OKeefe's theory found support among a number of researchers, there is no hard evidence to support the view that the Moon has been geologically active in the recent past. Some astronomers have witnessed discoloration of certain lunar craters—the so-called Lunar Transient Phenomenon (LTP)—which may be caused by degassing through fissures in the Moon's crust as it opens under gravitational tidal forces, but this reaction is a long way from full-blown volcanism. Nininger's lunar impact suggestion has the advantage of being dynamically feasible (six lunar "meteorites" have recently been found on Earth) but as noted earlier, studies of lunar material returned by the Apollo program does not support a lunar origin for tektites.

Terrestrial volcanism is not sufficiently powerful to distribute tektites over vast areas and could not produce the chemical composition of tektites. A terrestrial impact is the most obvious method, but strewn fields are not always associated with known craters. Those that are include the Czechoslovakian moldavites with the Nordlinger Ries crater in Germany, and the Ivory Coast tektites with Lake Bosumptwi, Ghana.

Appendix A

Glossary

Ablation: Process in which the surface of a meteoroid melts and vaporizes during its passage through the atmosphere.

Achondrite: Major classification of stony meteorite.

Aërolite: Obsolete term for a stony meteorite commonly used in the 19th century (Gk. *aer* = air; *lithos* = stone).

Aërolith: Obsolete term for a stony meteorite. Meteorite researchers were often called aërolithologists.

Alabandite: Mineral. Manganese sulfide.

Albite: Form of sodium-rich feldspar.

Amphoterite: Type of chondrite meteorite.

Andesine: Form of feldspar.

Angrite: Type of achondrite.

Anorthite: Form of calcium-rich feldspar.

Apollo–Amor–Aten Asteroids: Three groups of asteroids whose orbits either intersect that of the Earth, sometimes resulting in an impact, or approach it.

Aqueous Alteration: Alteration of minerals in a meteorite due to water-rich fluids circulating through the body.

Asteroid: Term suggested by Herschel for the minor planets.

Asteroid Belt: Collective term used to describe the system of minor planets in near-circular heliocentric orbits between Mars and Jupiter.

Astrobleme: A very large, ancient, meteorite scar.

Ataxite: Major classification of iron meteorite which lacks macroscopic structure.

Aubrite: Type of achondrite.

Augite: Form of pyroxene.

Australite: Tektite found in Southern Australia and Tasmania.

127

Barringer Crater: Alternative name for Meteor Crater, Arizona.

Billitonite: Name given to those tektites found on Billiton Island.

Black Stone: A rock forming an integral part of the Kaaba shrine at Mecca thought by some to be a meteorite.

Bloedite: A hydrous mineral.

Bottlestone: Name given to Moldavite tektites because of their bottle-green coloration.

Breccia: Rock composed of broken fragments.

Bronzite: Form of pyroxene.

Bytownite: Form of feldspar.

Calcite: Mineral. Calcium carbonate.

Carbonaceous chondrite: Major classification of stony meteorite.

Ceres: More correctly 1 Ceres; the first, and largest asteroid to be discovered.

Chassignite: Classification of stony meteorite thought to have originated on Mars.

Chladnite: Obsolete collective term for aubrites and diogenites.

Chlorapatite: Mineral. Calcium-chlorine phosphate.

Chondrite: Major classification of stony meteorite.

Chondrule: Small sphere found in some stony meteorites.

Chromite: Mineral. Iron-chromium oxide.

Chrysolite: Form of olivine.

Cliftonite: Mineral. Form of carbon.

Clinobronzite: Form of low-calcium clinopyroxene.

Clinoenstatite: Form of clinopyroxene.

Clinoferrosilite: Form of clinopyroxene.

Clinohypersthene: Form of clinopyroxene.

Chinopyroxene: Pyroxene of monoclinic crystal form.

Close textured crust: Type of fusion crust.

Coesite: Very high pressure polymorph of silica.

Cohenite: Mineral. Iron carbide.

Comet: Volatile-rich body which is usually in a highly eccentric heliocentric orbit.

Coone Butte: Alternative name for Meteor Crater, Arizona.

Cosmic Dust: Collective term for micrometeoroids.

Cosmic ray exposure age: The time a meteorite has existed as a meter-sized body in space.

Cosmic velocity: The velocity of a body in orbit around the sun.

Crater: Bowl-shaped cavity produced by a meteorite impact.

Craterlet: A small crater 10–100 cm diameter.

Crater pit: The smallest form of crater with a diameter of less than 10 cm.

Cristobalite: Very high temperature polymorph of silica.

Daubréelite: Mineral unknown in terrestrial rocks. Iron-chromium sulfide.

Deliquescent: The ability to readily absorb water.

Diogenite: Type of achondrite.

Diopet: A possible meteorite which was placed in the headdress of the statue of Diana at Ephesus.

Diopside: Form of Calcium-rich pyroxene.

Dispersion ellipse: Elliptical area over which meteorites may be found after a fall.

Dustball Model: Model of meteoroidal structure associated with comets.

Earth grazers: Synonymous with Apollo-Amor-Aten asteroids.

Ensisheim: Second oldest known meteorite fall from which material has been preserved. It fell on November 16, 1492.

Enstatite: Form of pyroxene.

Enstatite achondrite: Type of stony meteorite.

Enstatite chondrite: Type of stony meteorite.

Epsomite: Hydrous mineral. Magnesium sulfate.

Eucrite: Type of achondrite.

Explosive cratering: The formation of a crater by a large meteorite that vaporized on impact.

Fall: An observed meteorite event.

Farringtonite: Mineral found only in meteorites. Magnesium phosphate.

Fayalite: Iron-rich olivine.

Feldspar: One of the main types of alumino-silicate minerals.

Ferrohortonolite: Form of olivine.

Ferrosilite: Form of pyroxene.

Find: The discovery of a meteorite whose fall has not been observed.

Fireball: A brilliant meteor brighter than magnitude −4.0.

Forsterite: Magnesium-rich olivine.

Fusion crust: A thin, dull, or lustrous crust covering the surface of a meteorite.

Garnet: Mineral. Calcium-aluminum silicate.

Gehlenite: Mineral. Calcium silicate.

Giordano Bruno: A crater on the lunar surface whose formation may have been witnessed by a group of Canterbury monks on June 18 , 1178 and recorded by Gervase.

Grahamite: An obsolete classification of siderolite meteorite.

Graphite: Mineral. Carbon.

Ground track: The path of a fireball when projected onto the Earth's surface.

Gypsum: Mineral. Hydrated calcium sulfate.

Hadshar-al-Aswad: Synonymous with the Black Stone.

HED meteorites: Collective term for howardites, eucrites, and diogenites.

Hedenbergite: Form of pyroxene.

Hortonolite: Form of olivine.

Howardite: Type of stony meteorite.

Hyalosiderite: Form of olivine.

Hypersthene: Form of pyroxene.

Hypervelocity impact: The collision between two bodies at velocities of kilometers per second.

Ilmenite: Mineral. Iron-titanium oxide.

Impactite: Impact debris.

Ionization: Process by which a neutral atom becomes electrically charged. It is then called an ion.

Iron: Collective term for iron- and nickel-rich meteorites.

Isomorphism: Minerals that can have different chemical compositions but closely related formulae.

Knobby crust: Type of fusion crust.

Krypton: A theoretical Jovian-type planet which, according to M.W. Ovenden, disintegrated to form the asteroid belt.

Labradorite: Form of feldspar

L'Aigle: A small town in Normandy, France, the site of a historic meteorite shower on April 26, 1803, which convinced the French Academy of Sciences of the cosmic nature of meteorites.

Lawrencite: A meteoritic mineral not found on Earth. Iron chloride.

Lei-gong-mo: Inkstone of the Thundergod, the name given to certain types of tektite by the inhabitants of South East China.

Lodranite: Classification of stony-iron meteorite.

Lonsdaleite: Mineral. High pressure polymorph of carbon.

Luminous train: A column of ionized gases and dust which remain visible for a time after a meteor or fireball event.

Luminous trajectory: That part of the trajectory of a meteor or fireball in which ablation and ionization occur.

Magnesite: Mineral. Magnesium carbonate.

Magnetite: Mineral. Iron oxide.

Maskelynite: Glass of plagioclase composition, unknown on Earth.

Merrillite: Meteoritic mineral unknown on Earth. Sodium-calcium phosphate.

Mesosiderite: Classification of stony-iron meteorite.

Metabolite: An octahedrite meteorite in which the normally characteristic Widmanstätten Pattern has been thermally obliterated.

Metamorphism: Changes in textural crystallization, element distribution, and mineral abundance in a meteorite as a result of thermal and pressure changes.

Meteor: The light phenomena which result from entry into the atmosphere of a meteoroid or micrometeoroid.

Meteor Crater: A large and well-preserved crater in the Northern Arizona Desert.

Meteorite: A mineral or mineral aggregate of cosmic origin, with a diameter of between 1 mm and 1 km, that has impacted with a planet.

Meteoriticist: One who studies meteoritics.

Meteoritics: The study of rocks and minerals of cosmic origin, and their associated phenomena (e.g., meteors, fireballs, etc.)

Meteoritics: The Journal of the Meteoritical Society.

Meteoritic mineral: A mineral which occurs in meteorites but not in terrestrial rocks.

Meteoritophile: A lover of meteorites and tektites.

Meteorodes: Eroded meteorite fragments.

Meteoroid: A mineral or mineral aggregate of cosmic origin, with a diameter of between 1 mm and 1 km, which exists in the cosmic environment.

Micrometeorite: A mineral or mineral aggregate of cosmic origin, with diameter of less than 1 mm, that has impacted with a planet.

Micrometeoroid: A mineral or mineral aggregate of cosmic origin, with a diameter of less than 1mm, that exists in the cosmic environment.

Microtektite: A tektite with a diameter of less than 1 mm.

Minor planet: A solid, often irregularly-shaped rocky body, 1 to 100 km in diameter, in a heliocentric orbit. An asteroid.

Moissanite: Mineral. A silicon carbide found in a few meteorites but not on Earth.

Moldavite: Tektite found in the vicinity of the Moldau River, Czechoslovakia.

Moon stone: Term used in a few old chronicles to describe a meteorite obviously in the belief that they originated on the Moon.

Monomict breccia: A breccia in which the clasts are of the same material as the host.

Monosomatic: Composed of a single crystal.

Nakhlite: Classification of stony meteorite believed to have originated on Mars.

Nepheline: Mineral. Sodium-aluminum silicate.

Net crust: Type of fusion crust.

Neumann Lines: Very fine series of lines, some crossing one another, which are characteristic of mildly shocked kamacite.

Nickel-iron: Alloy. Principal component of the so-called iron meteorites.

Nininger Prize: Award established by Harvey and Addie Nininger to encourage an interest in meteoritics among students.

Octahedrite: Major classification of iron meteorite characterized by the Widmanstätten Pattern.

Oldhamite: Mineral. Calcium sulfide.

Oligoclase: Form of feldspar.

Olivine: One of the principal silicate minerals.

Olivine-bronzite chondrite: Major classification of stony meteorite.

Olivine-hypersthene chondrite: Major classification of stony meteorite.

Ordinary chondrite: The most abundant type of stony meteorite.

Oriented meteorite: A meteorite showing a distinct ablation pattern on the frontal surface caused by either a lack of rotation or rotation along the flight axis during its passage through the atmosphere.

Orthopyroxene: Pyroxene of orthorhombic crystal form.

Osbornite: A meteoritic mineral not found naturally on Earth. Titanium nitride.

Pallasite: Classification of stony-iron meteorite.

Parent body: The original bodies from which meteorites are derived.

Path: The apparent trajectory of a meteor or fireball when seen against the celestial sphere.

Pentlandite: Mineral. Nickel-iron sulfide.

Percussion crater: Small crater or impact pit.

Perryite: Mineral. Nickel-iron silicide

Piezoglypt: Synonymous with regmaglypt.

Pigeonite: Form of pyroxene.

Plagioclase: Form of feldspar.

Planetesimal: Literally "small planet." Thought to have existed in the early Solar System. As they coalesced they formed larger planet-size bodies.

Planetoid: Term sometimes used for asteroid.

Poikiloblastic: Metamorphic texture in which one crystal has grown to enclose others.

Point of retardation: Point at which the cosmic velocity of a meteoroid passing through the atmosphere is lost and the meteor, fireball and ablation phenomena cease.

Polymict breccia: Breccia in which the clasts are compositionally different from the host material.

Polymorphism: Phenomenon in which a compound may occur in more than one crystal form.

Polysomatic: Composed of several crystals.

Porous crust: Type of fusion crust.

Prairie Network: A network of sixteen automated camera stations that was operated by the Smithsonian Astrophysical Observatory between 1964–75 in the hope that photographs of bright fireballs would lead to the recovery of meteorites.

Pultusk Peas: Name given to approximately 100,000 pea-sized chondrites which fell at Pultusk, Poland on January 30, 1868.

Pyroxene: One of the principal groups of silicate minerals.

Pyrrhotite: Mineral. Iron sulfide.

Quartz: Mineral. Low-pressure polymorph of silica.

Queenstownite: Tektite found in the vicinity of Queenstown, Tasmania.

Regmaglyph: Synonymous with regmaglypt.

Regmaglypt: Thumb-like indentation on the rear and lateral surfaces of some meteorites.

Regolith: Layer of debris, mainly dust and broken rock fragments, which cover the surfaces of some planets and satellites and which is mainly the result of intensive meteorite bombardment in the early Solar System.

Reichenbach Lamellae: Thin plates which occur in regular patterns in some meteorites.

Rhabdite: Meteoritic mineral unknown on Earth. It is a form of schreibersite (a nickel-iron-cobalt phosphide).

Ribbed crust: Type of fusion crust.

Roedite: Obsolete classification of stony meteorite.

Rutile: Mineral. Titanium dioxide.

Schreibersite: A meteoritic mineral unknown on Earth. Nickel-iron-cobalt phosphide.

Scoriaceous crust: Type of fusion crust

Shatter cone: Fractured cone-shaped rock structure found beneath the floors of some large craters. Its presence indicates high impact pressures.

Shergottite: Classification of stony meteorite believed to have originated on Mars.

Shooting star: Layman's term for a meteor.

Shower: A display of meteors or a large, multiple meteorite fall.

Siderite: General term used to describe an iron meteorite regardless of composition or structure.

Siderolite: General term used to describe a stony-iron meteorite rich in silicates.

Siderophyre: Classification of stony-iron meteorite.

Silberschlag: A lunar crater named after Johann E. Silberschlag (1721-91), a German theologian who was the first person to calculate the orbit of a meteoroid.

Sodalite: Mineral. Sodium-aluminum-chlorine silicate.

Solid-solution series: A mineral composed of mixed crystals resulting from complete ionic substitution while in the solid state.

Spinel: Mineral. Magnesium-aluminum oxide.

Stishovite: Mineral. High-pressure polymorph of silica.

Stone: Shortened term for a stony meteorite.

Strewn field: An area in which a number of tektites or meteorites can be found.

Striated crust: Type of fusion crust.

Tail: The stream of ionized gases and dust released during the ablation of a meteoroid in the atmosphere.

Tektite: A highly siliceous, glassy body found in some parts of the world whose origin is unknown for certain but which is probably linked to large meteorite impacts.

Tesseral octahedrite: Type of octahedrite meteorite which shows orientation of lamellae according to the faces of a cube.

Thunderstone: Term popularly used in the seventeenth and eighteenth centuries to describe meteorites in the belief that they were terrestrial rocks that had been struck by lightning.

Train: A narrow column of ionized gases and dust which remains visible for a short time after a meteor or fireball event.

Trajectory: The real path of a meteor or fireball through the atmosphere.

Tridymite: Mineral. High-temperature polymorph of silica.

Troilite: Mineral. Iron sulfide.

Tunguska Event: A major explosion in the air above Siberia which occurred on June 30, 1908 and which is thought to have been the result of an asteroidal or cometary impact.

Ureilite: Classification of stony meteorite.

Vesta: Or 4 Vesta. The fourth asteroid to be discovered; believed to be the parent body of HED meteorites.

Warty crust: Type of fusion crust.

Widmanstätten Pattern: Series of intersecting plates which is characteristic of octahedrite meteorites.

Wollastonite: Form of pyroxene.

Appendix B

Meteorite and Tektite Collections

The following museums and universities have collections of meteorites and tektites which are well worth visiting.

Arizona

Arizona State University, Center for Meteorite Studies, Tempe, Arizona 85281

College of Mines, Department of Geology, University of Arizona, Tucson, Arizona 85721

California

California Division of Mines & Geology, Ferry Building, San Francisco, California

Griffith Observatory, Los Angeles, California

Institute of Geophysics & Planetary Sciences, University of California, Los Angeles, California 90024

Museum of the California Academy of Sciences, Golden Gate Park, San Francisco, California

Colorado

Denver Museum of Natural History, Denver, Colorado 80202

Connecticut

Peabody Museum of Natural History, Yale University, New Haven, Connecticut

Illinois

Field Museum of Natural History, Roosevelt Road at Lake Shore Drive, Chicago, Illinois 60605

Natural History Museum, University of Illinois, Urbana, Illinois

Massachusetts

Mineralogical Museum, Harvard University, 24 Oxford Street, Cambridge, Massachusetts 02138

Pratt Museum, Amherst College, Massachusetts 01002

Michigan

Natural Science Building, The University of Michigan, Ann Arbor, Michigan 48103

Missouri

Museum of Science & Natural History, Oak Knoll Park, St. Louis, Missouri 63105

New Mexico

Institute of Meteoritics, University of New Mexico, Albuquerque, New Mexico 87106

New York

American Museum of Natural History, Central Park West at 79th Street, New York, New York 10024

City University of New York, New York, New York

North Carolina

North Carolina Museum of Natural History, Salisbury Street, Raleigh, North Carolina 27601

Pennsylvania

Academy of Natural Sciences, 19th and The Parkway, Philadelphia, Pennsylvania 19103

Texas

Texas Christian University, Forth Worth, Texas

University of Texas, Austin, Texas 78710

Washington, D.C.

U.S. National Museum, Smithsonian Institution, Constitution Avenue at 10th Street, Washington, D.C. 20560

Wisconsin

Oshkosh Public Museum, 1331 Algoma Boulevard, Oshkosh, Wisconsin 54901

Argentina

Museo Bernardino Rivadavia, Avenida Angel Gallardo 470, Buenos Aires

Australia

Australian Museum, 6-8 College Street, Sudney South, New South Wales

Australian National University, Canberra, A.C.T.

South Australian Museum, North Terrace, Adelaide, South Australia

Western Australian Museum, Francis Street, Perth, Western Australia

Austria

Naturhistorisches Museum, Burgring 7, Vienna

Canada

Museum of the Geological Survey, 601 Booth Street, Ottawa

Denmark

Mineralogical Museum of the University, Oster Voldgade 5-7, 1350 Copenhagen

Eire

National Museum of Ireland, Kildare Street, Dublin

France

Collection de Mineralogie, Museum National d'Histoire Naturelle, 57 Rue Cuvier, Paris-5^{e}

Germany

Mineralogisches Institut der Universitat, Liebfrauenweg 3, Bonn

Italy

    Vatican Observatory, Castle Gandolfo, Vatican

Mexico

    Institute of Geology, R.Cipres, Mexico City

United Kingdom

    British Museum (Natural History), Cromwell Road, London SW7

Appendix C

Useful Addresses

C.1 Suppliers

The following companies supply meteorites and tektites, some also stock accessories such as display stands, chemicals, etc.

Armagh Planetarium, College Hill, Armagh, Northern Ireland, United Kingdom, BT61 9DB Tel: (0861) 524-725

Astral Projections, Box 24423, Little Rock, AR 72221 Tel: (501) 224-1789

Bethany Sciences, P.O. Box 3726, New Haven, CT 06525 Tel: (203) 393-3395

Blaine Reed, 907 County Road 207, No. 17, Durango, CO 81301 Tel: (303) 259-5326

Crystal Connection, 28 Avon St, Apt. 3, New Haven, CT 06511 Tel: (203) 777-4507

Cureton Mineral Co., Box 5761, Tucson, AZ 85703 Tel: (602) 743-7239

David New, Box 278, Anacortes, WA 98221 Tel: (206) 293-2255

Everything in the Universe, 5248 Lawton Avenue, Oakland, CA 94618 Tel: (415) 547-6523

Fireball Electronics, 246 East 52nd St., Odessa, TX (915) 366-4802

Mineralogical Research Co., 15840 East Alta Vista Way, San Jose, CA 95127 Tel: (408) 923-6800

Oklahoma Meteorite Lab., Box 1923, Stillwater, OK 74076 Tel: (405) 372-2311

R.A. Langheinrich, 326 Manor Avenue, Cranford, NJ 07016 Tel: (201) 276-2155

Robert A. Haag, Box 27527, Tucson, AZ 85726 Tel: (602) 882-8804

Science Graphics, Box 7516, Bend, OR 97708 Tel: (503) 389-5652

Star Magic, 36 East 23rd St., New York, NY 10021 Tel: (212) 794-8591

C.2 Organizations

Meteoritics falls uncomfortably on the border between astronomy and geology. As a result, both geological and astronomical bodies, such as the British Astronomical Association, have been reluctant to cater to those interested in meteorites, tektites, and planetary geology. There are, however, two organizations which enthusiasts may find beneficial to their interests:

Dr. Roger H. Hewins, Treasurer
Meteoritical Society
Department of Geological Sciences
Rutgers University
New Brunswick, NJ 08903

The Meteoritical Society is really a professional body but it does welcome amateurs. The Society's Journal, *Meteoritics*, carries mainly research papers but occasionally publishes articles of interest to non-specialists. *Meteoritics* also contains a supplement, *The Meteoritical Bulletin*, which gives details of new falls and finds and which serves to update the Catalogue of Meteorites.

Society of Meteoritophiles
9 Airedale
Hadrian Lodge West
Wallsend, Tyne & Wear
United Kingdom, NE28 8TL

The Society is a new organization aimed mainly at amateur collectors. At the time of writing (July 1990) the first issue of the Society's magazine, *Impact!*, is scheduled to appear in April 1991. *Impact!* will carry news

Fig. C-1 : *Eleven past Presidents of the Meteoritical Society. They are from left to right John A. Russell (1958–62), Peter M. Millman (62–66), Carleton B. Moore (66–68), Klaus Keil (68–70), John A. Wood (70-72), Robert Brett (72–74), Ursula B. Marvin (74–76), Paul Pellas (76–78), John T. Wasson (78–80), Vagn F. Buchwald (80-82) and George W. Wetherill (82–84). (Courtesy Dr. Peter M. Millman)*

of new falls and finds, details of recent developments, educational articles, advertisements from members wishing to trade or exchange specimens, and correspondence.

C.3 Nininger Award

A $1,000 prize is awarded annually to the student or students who have submitted the most meritorious essay on meteoritics to the Nininger Award Committee

The Award was established in 1961 by Harvey and Addie Nininger "to generate interest in meteoritics among the greatest number of student scientists." It is administered by the Center for Meteorite Studies at Arizona State University. It has often been shared between two and sometimes three students, so that in the first 21 years it was awarded 43 times. Some 59% of the recipients continue their studies of meteorites after leaving full time education. Details can be obtained from:

The Nininger Award
c/o The Director
Center for Meteorite Studies
Arizona State University
Tempe, AZ 85281

Appendix D

Bibliography

This bibliography is concerned mainly with books on meteorites but also contains some references to comets and asteroids.

Adams, P., *Moon, Mars and Meteorites* HMSO, London (1977).

Arnold, J.R., *The Origin of Meteorites as Small Bodies*, North Holland Publishing Co., Amsterdam (1964).

Barnes, V.E. & M.A., *Tektites: Benchmark Papers in Geology* Hutchinson & Ross, Stroudsberg (1973).

Barringer, D.M., *Meteor Crater (Formerly Called Coon Mountain or Coon Butte) in North Central Arizona* (1910).

Brandt, J.C. & Chapman, R.D., *Introduction to Comets*, Cambridge University Press (1981).

Buchwald, V.F., *Handbook of Iron Meteorites*, California University Press & Arizona State University (1975) [3 Vols.].

Clarke, R.S., *et al*, *The Allende, Mexico, Meteorite Shower*, Smithsonian Inst. Press, Washington D.C. (1971).

Cohen, E., *Meteoritenkunde Vols. 1-3* Stuttgart, (1894, 1903 & 1905).

Cunningham, C.J., *Introduction to Asteroids*, Willmann-Bell, Richmond, Virginia (1988).

Davies, J.K., *Cosmic Impact*, Fourth Estate (1986).

Delsemme, A.H. (Ed.), *Comets, Asteroids, Meteorites: Interrelationships, Evolution and Origins*, Toledo University Press (1977).

Dodd, R.T., *Meteorites: A Petrologic-Chemical Synthesis*, Cambridge University Press (1981).

Dodd, R.T., *Thunderstones and Shooting Stars*, Harvard University Press (1986).

Farrington, O.C., *Meteorites*, (1915).

Fessenkov, V.G., *The Sikhote-Alin Iron Meteorite Shower* Acad. Science USSR, Moscow (1959).

Flight, W., *A Chapter in the History of Meteorites* (1887).

Foster, G.E., *The Barringer (Arizona) Crater* (1957).

Gehrels, T., *Asteroids*, Arizona University Press (1979).

Geiss, J. & Goldbery, E.D. (Eds.), *Earth Sciences and Meteorites*, North Holland Publishing Co., Amsterdam (1963).

Gomes, C.B. & Keil, K., *Brazilian Stone Meteroites*, University New Mexico Press (1988).

Graham, A., Bevan, A.W.R. & Hutchison, R. *Catalogue of Meteorites*, [4th Edition], University of Arizona Press, Tucson, and the British Museum (Natural History), London (1985).

Hawkins, G.S., *Meteors, Comets and Meteorites*, McGraw-Hill, New York (1964).

Heide, F., *Meteorites*, Chicago University Press (1964).

Hoyt, W.G., *Coon Mountain Controversies*, Arizona University Press, (1987).

Hsu, K.J. *The Great Dying*, Harcourt Brace Jovanovich (1986).

Hutchison, R., *The Search For Our Beginning*, Oxford University Press (1983).

King, E.A. (Ed.), *Chondrules and their Origin*, Lunar & Planetary Inst. (1983).

Krinov, E.L., *Giant Meteorites*, Pergamon Press, London (1962).

Krinov, E.L., *The Tunguska Meteorite*, Acad. Sciences USSR, Moscow (1949).

Lancaster-Brown, P., *Comets, Meteorites and Man*, Robert Hale, London (1974).

La Paz, L., *Topics in Meteoritics*, University New Mexico Press, (1969).

Lockyer, J.N., *The Meteoritic Hypothesis*, (1890).

Lyttleton, R.A., *The Comets and their Origins*, Cambridge University Press (1982).

Mason, B., *Handbook of Elemental Abundances in Meteorites*, Gordon & Breach, N.Y. (1971).

Mason, B., *Meteorites*, J.Wiley, N.Y. (1962).

Mark, K., *Meteorite Craters*, Arizona University Press (1987).

McCall, J.G.H., *Astroblems—Cryptoexplosion Structures*, Hutchinson & Ross, N.Y. (1979).

McCall, G.J.H., *Meteorites and their Origins*, David & Charles, London (1973).

McSween, H.Y., *Meteorites and their Parent Planets*, Cambridge University Press (1987).

Middlehurst, B.M. & Kuiper, G.P., *The Moon, Meteorites and Comets*, Chicago University Press (1963).

Millman, P.M. (Ed.), *Meteorite Research*, D.Reidel, Dodrecht (1969).

Moore, C.B. (Ed.), *Researches on Meteorites*, J.Wiley, N.Y. (1962).

Nininger, H.H., *Ask A Question About Meteorites*, (1961).

Nininger, H.H., *Arizona's Meteorite Crater: Past, Present, Future*, World Press, Denver (1956).

Nininger, H.H., *Chips From The Moon*, Desert Press, El Centro, CA (1947).

Nininger, H.H., *A Comet Strikes The Earth*, American Meteorite, Lab., Denver (1942).

Nininger, H.H., *Find A Falling Star*, P.S. Eriksson, N.Y. (1972).

Nininger, H.H., *Our Stone-Pelt Planet*, Houghton-Mifflin, N.Y. (1933).

Nininger, H.H., *Out Of The Sky*, Dover Publishing (1959).

Nininger, H.H., & Huss, G.I. *The Nininger Collection of Meteorites*, Winslow, Arizona (1950).

O'Keefe, J.A., *Tektites and their Origin*, Elsevier (1976).

Paneth, F.A., *The Origin of Meteorites*, Oxford University Press (1940).

Porter, J.G., *Comets and Meteor Streams*, Chapman & Hall, London (1952).

Povenmire, H.R., *Fireballs, Meteors and Meteorites*, JSB Enterprises (1981).

Ramdohr, P., *The Opaque Minerals in Stony Meteorites*, Akademie-Verlag (1972).

Roddy, D.J., Pepin, R.O. & Merrill, R.B. *Impact and Explosion Cratering*, Pergamon (1977).

Sagan, C. & Druyan, A., *Comet*, M. Joseph, London (1985).

Sears, D.W. *The Nature and Origin of Meteorites*, Adam Hilger, Bristol (1978).

Sears, D.W., *Thunderstones: A Study of Meteorites Based on Falls and Finds in Arkansas*, Arkansas University Press (1988).

Silver, L.T., *Geological Implications of Impacts of Large Asteroids and Comets on the Earth*, Geological Society of America (1982).

Wasson, J.T., *Meteorites—Classification and Properties*, Springer Verlag (1974).

Wasson, J.T., *Meteorites: Their Record of Solar System History*, W.H. Freeman (1985).

Watson, F.J., *Between the Planets*, Harvard University Press (1956).

Whipple, F.L., *The Mystery of Comets*, Cambridge University Press, (1985).

Wilkening, L. (Ed.), *Comets*, Arizona University Press (1982).

Willey, R.R., *The Tucson Meteorites*, Smithsonian Inst. Press, Washington D.C. (1987).

Wood, J.A., *Meteorites and the Origin of Planets*, McGraw-Hill, (1968).

Appendix E

Further Readings

The history of meteorites is discussed in some detail by John Burke in his *Cosmic Debris* whilst Harvey Nininger's autobiography, *Find a Falling Star*, was published in 1972.

During the past three decades a number of books have dealt with individual meteorites and meteorite theory. Brian Mason's *Meteorites* is regarded as a classic, and Heide's book of the same title is still the best little volume ever to be published on the subject even though it is now over 30 years old and somewhat dated. John Wood's *Meteorites and the Origin of Planets* (1968) describes how the study of meteorites has helped to clarify our understanding of the origin of the Solar System. G.J.H. McCall gave an updated account of meteorites in *Meteorites and their Origins* (1973), and ten years later Robert Hutchison examined meteorites in a much broader context when he embarked on *The Search for our Beginning*. More recently, popular texts on meteorites have been published by Dodd in *Thunderstones and Shooting Stars*, and McSween in *Meteorites and their Parent Planets*. For anyone with a special interest in irons Vagn Buchwald's *Handbook of Iron Meteorites* is invaluable, and no genuine meteoritophile would be without a copy of the fourth edition of the *Catalogue of Meteorites* compiled by Graham Bevan and Hutchison.

Two excellent accounts of craters and cratering can be found in Kathleen Mark's *Meteorite Craters* and Hoyt's *Coon Mountain Controversies*.

Meteorites' parent bodies, asteroids, are covered in Cunningham's *Introduction to Asteroids*. Those who prefer a cometary origin for certain classes of meteorites might like to read The *Mystery of Comets* by the foremost authority on the subject, Fred Whipple, and the more technical treatise, *Introduction to Comets*, by Brandt and Chapman.

Relatively few books have been published on tektites, the most notable of which was O'Keefe's *Tektites and their Origin*.

Meteoritics, the journal of the Meteoritical Society, carries papers on all aspects of meteorite research. Although aimed mainly at the professional, the journal publishes a number of articles each year of interest to the non-specialist.

Details of all these publications can be found in the Bibliography.

General Index

Italics refer to illustrations.
Bold Print refers to tektites.

A

Ablation, 22, 29, 47 54
Achondrites, 13, *15*, 58 73ff, 77, 81, 85ff
Advertising, 12
Aëroliths, 73
Ages of meteorites, 45, 85, 88
Ages of tektites, **115ff**
Albedo, 70, 80
Alkanes, 82
Altitude, 19
American Meteorite Laboratory, 17
Amino Acid, 82
Amphoterites, 13, 40, 73ff, *76, 77,* 81
Angrites, 74ff, 85
Antarctica, 53, 89, 98
Aqueous Alteration, 75, 82
Aromatic Hydrocarbons, 82
Asteroids, 3, 7, 8 17 84, 87ff
Astroblemes, 26
Ataxites, 13, 46, 72, 74ff, 93, 104
Aubrites, 13, 40, 52, 74ff, 77, 85ff

B

Black Chondrites, 38ff, 80
Breccias, 58ff, 86, 87, 89, 90ff
Bustites, 73

C

Calcium, 40, 73ff, 84, 87
Calcium- & Aluminum-rich inclusions
    - *see* Inclusions
Camera networks, 19ff
Carbonaceous chondrites, 13, 39ff, 43, 45, 58, *59*, 63, *63*, 70, 72, 74ff, *76, 77,* 82ff, 108
Carboxylic acid, 82

Chassigny, 45, 72, 74, 85, 87ff
Chemicals, 11
Chladnites, 72
Chondrites, 13, *15*, 29, 38, 40, 42, 46, 52, 63, 64, 70, 72ff, 77, 108
Chondrules, 45, 58ff, *59, 61, 66,* 73ff, 79, 82, 89, 97, 101
Classification, 14, 71ff
Clasts, 58, 80
Cleavage planes, 111
Collecting, 11ff, 26, 32, 112ff, 123ff
Collections, 4ff, 25, 58, 71ff, 136ff
Color, 49, 86, 97, 100, 103, 111, 116ff
Comets, *6,* 8, 17, 84
Common Chondrites - *see* Ordinary chondrites
Composition, 33ff, 58, **122**
Cosmic Dust, 8
Cosmic ray exposure ages, 45ff, *46*
Cosmic velocity, 27
Cosmogenic nuclides, 45
Craters & cratering, 18, 22, 26ff, *28,* 96, 115, 116, 125
Cutting, 39, 57, 64, 92, 102

D

Damage, 22, 96, *97,* 98
Deep sea sediments, 114
Deliquescence, 99ff
Diogenites, 1, 13, *46,* 70, 72, 74, 77, 85ff
Dispersion ellipse, 17, *18*
Display methods, 11, 101ff, 105, 106, **124**
Dust train, 54

E

Ejecta, 30
Energy, 17
Enstatite achondrites - *see* Aubrites

Enstatite chondrites, 13, 40, 42, 75ff, *76*, *77*, 80, *80*
Etching, 11, 65, 90ff, 102ff
Eucrites, 13, 40, 46, *46, 50*, 52, 70, 72, 74, *77, 83*, 85, 86ff
Evaporitic deposits, 53
Exsolution, 42, 81
External features, 49ff

F

Falls, 9, 14ff, 25, *26*, 89
Ferric chloride, 43, 99
Ferrous chloride, 43, 99
Files, 14
Finds, 15, 24ff
Fireballs, 3, 5, 8, 15ff, 18ff, *21, 22*, 29
Flanges, **114**
Florida Fireball Patrol, 19
Flow line, 96
Fremdlinge, 63
Fusion crust, 16, 22, 29, 49ff, *50, 56*, 80, *83*, 86, 96, 108, **108**

G

Gallium, 73, 91
Gases, 45, 88
Germanium, 73, 91
Glass, 38, 58, 81, 82, **114**
Grahamite, 72
Groundtrack, 20

H

H-Chondrites - *see* Olivine-bronzite chondrites
Heat treatment, 65, 104
HED suite, 74, 89
Hexahedrites, 13, 40, 46, *64*, 65, 72, 74, *77*, 80, 92ff, 104
History, 1ff, **113ff**
Howardites, 13, 40, *46*, 52, 72, 74, 85ff
Hydrated sulfates, 37, 43

I

Identification, 96
Impact glass, 13
Impactites, 11, 32
Inclusions, 17, 45, 63, *63*, 72, 80, 84, **122**
Institute of Meteoritics, 9
Insurance, 13, 111
Internal features, 57,**122**

Ionization, 29
Iridium, 73, 91
Iron hydroxide, 43, 99
Iron sulfide, 4, 82
Isomorphism, 33ff
Isotopes, 43ff, 64
Isotropism, 40, 111

J

Jetting, 30
Jupiter, 5

L

L-Chondrites - *see* Olivine-hypersthene chondrites
Lattice diffusion, 45
Legal ownership, 32
Litholites, 73
Lithosiderites, 73
Lodranites, 13, 74, *77*, 89
Lunar Transient Phenomena (LTP), 125

M

Magnesium, 33, 40
Mars, 5, 45, 88
Mass loss, 29, 47
Mass spectrometer, 43
Mercury, 18, *28*, 80
Mesosiderites, 13, 40, 72ff, *77*, 89, *90*
Metamorphism, 75
"Meteoritic cancer", 99
Meteoritical Bulletin, 111
Meteoritical Society, 12, 142, *143*
Meteoritics, 9, 111
Meteorodes, 13, 32
Meteoroids, *6*, 27
Meteors, 5, *6*, 29
Meteor showers 8, 17, 84
Microscopy, 11, 57, 81, 94, 109ff, 122
Microtektites, 114
Minerals & mineral groups, 49, 64
    agate, 2
    akermanite, 85
    alabandite, 37
    albite, 35, 40, *41*
    aluminum, 40
    amphibole, 111
    andesine, 35, 40
    anorthite, 35, 40, *41*, 87
    augite, 36, 40, 59, 72ff, 85, 87

Minerals & mineral groups, *continued*
 bloedite, 37, 43
 bronzite, 36, 40, 72, 81, 87, 89
 bytownite, 35, 40
 calcite, 37
 carbides, 37, 42
 carbon, 33ff, *38, 39,* 42, 70, 82,
  86, 93
 carbonates, 37
 chlorapatite, 37, 43
 chloride, 37
 chromite, 35, 40, 42, 70, 87, 93
 chromium, 93
 chrysolite, 34, 36, 40
 cliftonite, 35, 39
 clinobronzite, 36, 81
 clinoenstatite, 36, 82
 clinoferrosilite, 36
 clinohypersthene, 36
 clinopyroxene, 36, 38, 40ff, 81, 99
 cobalt, 93
 coesite, 32, 35, 40
 cohenite, 37, 42, 93
 copper, 33, 35, 39
 cristobalite, 35, 40, 81
 daubréelite, 37, 40, *64*
 diamond, 35, 38ff, *38,* 86, 93
 diopside, 36, 40, 73, 81, 88
 dolomite, 37
 enstatite, 36, 40, 72, 80, *80,* 82,
  86
 epsomite, 37, 43, 53, 84
 farringtonite, 37, 43, 91
 fayalite, 33, 36, 40, 89
 feldspar, 35, 40, *41,* 72, 81, 108,
  111
 feldspathoids, 84
 ferrohortonolite, 36
 ferrosilite, 36, 89
 forsterite, 33, 36, 40, 82
 garnet, 36
 gehlenite, 36, 85
 gold, 39
 graphite, 13, 35, 38ff, *38,* 79, 86,
  93, 108
 gypsum, 37, 40, 43
 hedenbergite, 36
 hortonolite, 36, 40
 hyalosiderite, 36, 40
 hydromagnesite, 53
 hydrous phyllosilicates, 82
 hypersthene, 36, 40, 81, 87, 89
 ilmenite, 35, 87

Minerals & mineral groups, *continued*
 iron, 1ff, 12, 13, 15, *15,* 16, 33,
  39, 42ff, 46, 51, 52, 54, 57,
  64, 71, 74, 77, 91ff, 96, 99,
  102
 kaersutite, 87
 kamacite, 35, 39, 65, *66, 80,* 81,
  85, 92, 93, 108
 labradorite, 35, 40
 lawrencite, 37, 43, 99, 104
 limonite, 43, 99
 lonsdaleite, 35, 39
 magnesite, 37
 magnetite, 35, 40, 84, 87
 majorite, 38
 maskelynite, 36, 38, 40, 87, 111
 melilite, 85
 merrillite, 37, 42
 moissanite, 37, 42
 nepheline, 36
 nesquehonite, 53
 nickel, *69,* 73, 91, 93
 nickel-iron alloys see also
  kamacite and taenite, 35,
  39, *41,* 72, 81, 86, 87, 89,
  91, *92,* 97, 101
 nitride, 37
 obsidian, **113, 123, 145**
 oldhamite, 37, 40, 43
 oligoclase, 35, 40, 81
 olivine, 4, 32, 33, 36, 38, 40ff, *41,*
  62, 72, 73, 79, 81ff, 89ff, *92,*
  106, 108, 109
 orthoclase, 40, *41*
 orthopyroxene, 36, 38, 40, 72, 81,
  91
 osbornite, 37
 oxides, 35
 pentlandite, 37
 perovskite, 36, 85
 perryite, 36
 phosphates, 37
 phosphides, 37, 42
 pigeonite, 36, 40, 86, 87, 89
 plagioclase, 35, 38, 40, 72, *80,* 86,
  87, 89
 platinum, 63
 plessite, 65, 94
 pyroxene, 33, 36, 38, 40, 79, 82,
  85ff, 108
 pyrrhotite, 37, 42
 quartz, 33, 35, 40, 81
 rhabdite, 37, 42, 93

Minerals & mineral groups, *continued*
 ringwoodite, 38
 rutile, 35
 schreibersite, 37, 42, *66*, 72, 91,
  93, 108
 serpentines, 36, 82, 84
 silica, 40, 81
 silicates, 35ff, 40, 89
 silicide, 37
 sodalite, 36, 84
 spinel, 35, 82, 85
 starkeyite, 53
 stishovite, 32, 35, 40
 sulfates, 82
 sulfides, 37, 42, 79, 80
 taenite, 39, 65, *66, 69*, 81, 93, 94,
  108
 tridymite, 35, 40, 81, 87, 89
 troilite, 4, 37, 38, 42, 47, 54, *64*,
  69 72, 80, *80*, 81, 87, 91, 93,
  108
 wollastonite, 36
Monomict breccias, 58, 85, 86
Monosomatic chondrules, 59
Moon, 18, 30, 88, **114, 124**
Multiple falls, 16

N

Nakhlites, 40 45, 52, 74, 85, 88
Neumann Lines, 64ff, *64*, 67, 72, 92,
  101, 109

O

Octahedrites, 13, 42, 46, 65, *66*, 72, 74,
  75, *77*, 81, 90, 91, 93, 101,
  104
Odour, 47
Olivine-bronzite chondrites, 13, 20, *46,*
  51, 74ff, *76*, 81, 97
Olivine-hypersthene chondrites, 13, *46*,
  74ff, *76, 77*, 81
Orbits, 8, 20, 88
Ordinary chondrites, 40, 58, 63, 75
Organic compounds, 12, 82, **124**
Oriented meteorites, 51, 56, *57*, 96
Origin, 5ff, 88, **114ff. 124ff**

P

Pairing, 25
Pallasites, 12, 13, 32, 40, *41*, 72, 74,
  75, *77*, 91, *92*, 99, 106, *106*

Parent bodies, 39, 45, 65, 84, 86ff, 90
Petrologic types, 75
Phosphorus, 93
Photography, 19ff, 112
Photosensitivity, 100, 105
Pleochroism, 109, 110
Polarized light, 40, 109, **136**
Polishing, 11, 39, 57, 64ff, 90ff, 101ff
Polymict breccias, 58, 89
Polymorphism, 34ff, 81
Polysomatic chondrules, 59
Porosity, 79, 98, 108
Postal services, 13
Potassium, 40
Prairie Camera Network, 20
Preservation, 24, 86, 98ff **124**
Pressure, 40, 49, 79
Prices, 12, 13, **123, 124**

R

Radiometric dating, 43ff
Recovery, 16, 18ff
Refractory minerals, 63
Regmaglypts, 54, *55*, 96
Regolith breccias, 58, 80, 86
Reichenbach Lamellae, 69
Retardation point, 29
Rust, 51, 96, 99, 108

S

Scattering ellipse - *see* Dispersion
  ellipse
Scientific Event Alert Network
  (SEAN), 19
Shalkites, 72
Shatter cones, 11, 30
Shergottites, 45, 74, 85, 87ff
Shock, 29ff, 38ff, 45, 58, 65, 81, 88ff
Siderites, 73
Siderophyre, 72, 74, 75, 89, 91
Smithsonian Institution &
  Astrophysical Obervatory,
  8, *19*, 20, 32
Smoke train, 49
SNC suite, 45, 74, 87
Society of Meteoritophiles, 142
Society for Research on Meteorites, 8,
  9
Sodium, 40
Solid-solution series, 34, 40, 72, 85
Spallation, 45

Stones, 1, 2, 13, 15, 16, 29, 40, 46, 51,
     55, 58, 71, 74ff, 99, 101
Stony-irons, 13, *15*, 69, 74, 89ff
Strewn fields, 114ff, **115**, **125**
Sulfur, 35, 47, 80, 82, 93
Supernova, 45, 64
Swarms, 17

T

Tektites, **11**, **113ff**
Temperature, 16, 27, 30, 40, 49, 58, 65,
     82
Thunderstones, 4
Titius-Bode Law, 5
Tools and weapons, 1ff
Trading and traders, 9, 11ff, 30, 101,
     104, 141ff
Trajectory, 17, 19, 20
Tumbled irons, 13
Twinning, 111

U

Ureilites, 13, 39, 40, 85, 86, 74, 75, 77

V

Vaporization, 30
Velocity, 16, 27
Venus, 80
Vesta, 87

W

Water, 84, 87
Weathering, 51ff, 65, 96, 98
Widmanstätten Structure, 39, 42, 65ff,
     66, 67, 68, 69, 72, 90ff, *92*,
     96, 99, 101, 108, 109

Meteorite Index

Italics refer to illustrations.

A

Aegospotami, 1
Akyumak, *67*
Albareto, 4, 42
Allan Hills, 24, 88
Allende, *59, 63*, 84, 108
Angra dos Reis, 85
Antarctic Meteorites, 24
Aubres, 75, 86

B

Bayard, *52*
Beaver, 24
Bennett Co., *64*
Bjurböle, 80
Black Stone, 2
Braunau, 16
Brenham, 32, *92*
Bustee, 40

C

Camel Donga, *18*
Chassigny, 75, 88, 89
Clovis, *60, 61*
Coahuila, 42
Cosby's Creek, 39
Crab Orchard, *90*

D

Diopet, 2

E

Edmonton, *66*
Elephant Moraine, 24, 88
Ensisheim, 3

G

Governador, 88
Gujargaon, 47, *97*

H

Hadshar al Aswad, 2
Haig, *55*
Homestead, 18
Hraschina, 4

I

Innisfree, 20
Ivuna, 84

J

Jerslev. *66*
Johnstown, 17, *18*

K

Krasnojarsk, 4, 65
Kunashak, *18*

L

Lafayette, 88
L'Aigle, 5
La Lande, 79
Lodran, 75, 89
Lost City, 20, *22, 23*

M

Magura, 42
Middlesbrough, 29
Mighei, 84

N

Nakhla, 75, 88
Ngawi, *60, 61*
Nogata, 3
Novo-Urei, 39, 75
Neuvo Mercurio, *51, 56*

O

Odessa, *52*, 107, 108
Ornans, 84

P

Pasamonte, *50, 83*
Phrygia, 2
Pribram, 8
Pultusk, 18

R

Roosevelt Co., 25

S

Shergotty, 75, 88
Springwater, 43
Steinbach, 91

T

Tábor, 4

V

Valadares, 88
Vigarano, 84

W

Willamette, 24

W

Yamato Mts., 24
Yamato 791493, 89

Z

Zagami, 88

Name Index

Italics refer to illustrations.

A

Anaxagoras, 1
Aristotle, 3, 5

B

Berwerth, F. 114
Bhandari, H. 47
Biot, J.-B. 5
Boer, K. De, 124
Boisse, A. 71
Brett, R. *143*
Brezina, A. 73
Buchwald, V.F. *143*

C

Chladni, E.F.F. 4
Clayton, R. 45
Cohen, E. 8

D

Darwin, C. 113, 123
Daubrée, G.-A. 37, 73
David, T.W.E. 124
Diogenes, 1, 75
DuFresne, E.R. 43

E

Easton, E.W. 124

F

Farrington, O.C. 8, 37
Floran, R.J. 90

G

Goldschmidt, V. 124

H

Haidinger, W. 42
Hauser, E. 16
Heyman, D. 58
Huss, G.I. 17
Hutchison, R. 17

J

Jain, A.V. 58
Jehangir, 2
Jerofejeff, N. 39
Joule, J. 5

K

Keil, K. *143*
Khan, M.A.R. 124
Konig, K. 5
Krinov, E.L. 17, 53

L

Latchinoff, P.A. 39
Lawrence Smith, J. 37, 42
Leonard, F.C. 8
Lipschutz, M.E. 58
Liu Sun, 113
Livy, 1
Lovering, J.F. 73

M

Marvin, U.B. *143*
Mason, B. 91
McCall, G.J.H. 9
Medvedyev, 4
Merrill, G.P. 8, 17, 37, 43
Meunier, S. 73
Michel, H. 124
Millman, P.M. *143*
Mohammed, 2
Mohs, F. 5
Moissan, H. 37, 42
Moore, C.B. *143*

Mu Tuan Lin, 1

N

Neumann, J.G. 65
Nininger, H.H. 8, 30, 31, 114, 124, 125
Numa Pompilius, 2

O

O'Keefe, J.A. 114, 125
Olivier, C.P. 17

P

Pallas, P.S. 4, 75
Partsch, P. 5, 71
Pellas, P. *143*
Piazzi, Fr. G. 5
Pliny, 1
Plutarch, 1
Prior, G. 73

R

Reichenbach, C. von, 72
Rose, G. 58, 72
Roy, S.R. 43
Russell, J.A. *143*

S

Schiaparelli, G.V. 8
Schmus, W.R. van, 73, 75
Schreibers, C. von, 5 37
Shephard, C.U., 71, 72, 73
Simashko, Ju. I. 8
Spencer, L.J. 124
Stepling, J. 4
Storey-Maskelyne, N. 5, 37, 40, 73
Streich, V. 114
Stütz, B. 4
Suess, F.E. 114
Swartz, G. 20

T

Taylor, G.J. 58
Thomson, G. 65
Troili, Fr. D. 4, 42
Tschermak, G. 42, 72

W

Wasson, J.T. 17, 73, *143*
Wetherill, G.W. *143*
Wherry, E.T. 42
Widmanstätten, Count A. de, 65
Wood, J.A. 73, 75, *143*
Woolaston, W.H. 5